Studies in Computational Intelligence

Volume 1023

The series “Studies in Computational Intelligence” (SCI) publishes new developments and advances in the various areas of computational intelligence—quickly and with a high quality. The intent is to cover the theory, applications, and design methods of computational intelligence, as embedded in the fields of engineering, computer science, physics and life sciences, as well as the methodologies behind them. The series contains monographs, lecture notes and edited volumes in computational intelligence spanning the areas of neural networks, connectionist systems, genetic algorithms, evolutionary computation, artificial intelligence, cellular automata, self-organizing systems, soft computing, fuzzy systems, and hybrid intelligent systems. Of particular value to both the contributors and the readership are the short publication timeframe and the world-wide distribution, which enable both wide and rapid dissemination of research output.

Indexed by SCOPUS, DBLP, WTI Frankfurt eG, zbMATH, SCImago.

All books published in the series are submitted for consideration in Web of Science.

More information about this series at https://link.springer.com/bookseries/7092

Victor Chang · Harleen Kaur · Simon James Fong
Editors

Artificial Intelligence and Machine Learning Methods in COVID-19 and Related Health Diseases

Springer

Editors
Victor Chang
Department of Operations and Information Management
Aston University, Aston Business School
Birmingham, UK

Harleen Kaur
Department of Computer Science and Engineering
Jamia Hamdard University
New Delhi, India

Simon James Fong
Department of Computer and Information Science
University of Macau
Macau, China

ISSN 1860-949X ISSN 1860-9503 (electronic)
Studies in Computational Intelligence
ISBN 978-3-031-04596-7 ISBN 978-3-031-04597-4 (eBook)
https://doi.org/10.1007/978-3-031-04597-4

This Springer imprint is published by the registered company Springer Nature Switzerland AG
The registered company address is: Gewerbestrasse 11, 6330 Cham, Switzerland

Preface

The most severe issue that concerns the world during this period is the outbreak of the novel coronavirus (COVID-19). The rapid spread of the virus around the world poses a real threat to all countries. As a result, researchers must pay attention to studying the details of this calamity. COVID-19 symptoms may be similar to other viral chest diseases. Some of the symptoms may cause the doctor's uncertainty in making the correct diagnosis decision due to the novelty of this virus. The recent diagnosis of COVID-19 is based on real-time reverse-transcriptase polymerase chain reaction (RT-PCR) and is regarded as the gold standard for confirmation of infection. It has already been widely recognized that deep learning techniques can potentially have a substantial role in streamlining and accelerating the diagnosis of COVID-19 patients. Over the past weeks, numerous open dataset enterprises have been set up to aid the researchers in developing and improving methods that could contribute to countering the Corona pandemic. Various techniques need to be developed to report the above unique problems in the diagnosis of COVID-19. This book focuses on novel analysis techniques related to COVID-19.

This Springer book provides a perfect platform to submit chapters that discuss the prospective developments and innovative ideas in artificial intelligence and machine learning techniques in the diagnosis of COVID-19.

COVID-19 is a huge challenge to humanity and the medical sciences. So far as of today, we have been unable to find a medical solution (vaccine). However, globally we are still managing the use of technology for our work, communications, analytics, and predictions with the use of advancement in data science, communication technologies (5G & Internet), and AI. Therefore, we might be able to continue and live safely with the use of research in advancements in data science, AI, machine learning, mobile apps, etc., until we can find a medical solution such as a vaccine.

There are urgent needs globally to understand how to tackle this challenge. In terms of computing and multimedia research, scientists can offer insights, recommendations, and new discoveries, which may offer positive impacts and findings related to the causes, cure, and analysis of treatment. The recent diagnosis of COVID-19 is based on real-time reverse-transcriptase polymerase chain reaction (RT-PCR) and is regarded as the gold standard for confirmation of infection. It has already been widely recognized that advanced AI and data science techniques can potentially have a substantial role in streamlining and accelerating the diagnosis of COVID-19 patients, offering high-quality research outputs and accurate predictive modeling. Therefore, this requires pioneering methods such as deep learning, artificial intelligence, and computational intelligence since they are highly important.

Together with innovative multimedia techniques, innovative AI and data science for COVID-19 can provide added value for scientists. In this book/book chapter, we seek high-quality and unpublished work based on pioneering AI, data science, and multimedia techniques and findings.

During this COVID-19 crisis, there have been a number of privacy and security issues of personal details. This can be securely managed with smart contracts in intelligent technologies using pioneering AI and data science techniques.

This Springer book will be useful for readers and researchers to apply techniques, methods, algorithms, and application of AI and data science methods/techniques for further advancements of research.

We have selected 11 chapters after the vigorous review process. Each chapter has demonstrated the research contributions and research novelty. Each group of authors must fulfill the following requirements:

- Significance to the readership of the journal
- Relevance to the special issue
- Originality of idea, technical contribution, and significance of the presented results
- Quality, clarity, and readability of the written text
- Quality of references and related work

We are delighted to have these 11 chapters to make our book project a high-quality scholarly project for research communities.

Birmingham, UK
New Delhi, India
Macau, China
December 2021

Prof. Victor Chang
Dr. Harleen Kaur
Dr. Simon James Fong

Contents

Uncertainty Propagation and Salient Features Maps in Deep Learning Architectures for Supporting Covid-19 Diagnosis 1
Iñaki Rodríguez-García, Tomás Sánchez-Pastor, Julia Vázquez-Escobar, Juan Luis Gómez-González, and Miguel Cárdenas-Montes

A Review of Machine Learning Techniques to Detect and Treat COVID-19 Using EHR Data 23
Sagnik Nayak, Chirantan Ganguly, and Anil Kumar Gupta

Machine Learning-Based Emerging Technologies in the Post Pandemic Scenario 51
R. L. Priya, A. Abirami, and Naman Desai

Covid-19 Face Mask Detection Using Deep Learning Techniques 91
T. P. Pushpavathi and Siranjeevi Narasimman

Computational Intelligence in Identification of Some FDA Approved Drug Compounds for Treatment of COVID-19 109
Debanjan Mitra and Pradeep K. Das Mohapatra

Biomedical Data Driven COVID-19 Prediction Using Machine Learning Approach 123
Arijit Ghosal, Nikita Gupta, Esha Nandi, and Harshita Somolu

Vaccination, Lock-Down, Measures and Time-Varying Reproduction Number Based Estimation and Control of COVID-19 Outbreak in Turkey 139
Meric Cetin and Selami Beyhan

Diabetes Analysis with a Dataset Using Machine Learning 161
Victor Chang, Saiteja Javvaji, Qianwen Ariel Xu, Karl Hall, and Steven Guan

Impact of Covid-19 on Indian Agriculture 189
Debabrata Mondal

COVID-19 Patients Management and Triaging Using Machine Learning Techniques 211
Harleen Kaur, Iftikhar Alam, Ritu Chauhan, and Bhavya Alankar

Coordination of Covid-19 Vaccation: An Optimization Problem and Related Tools Derived from Telecommunications Systems 227
Christophe Gaie and Markus Mueck

Correction to: Uncertainty Propagation and Salient Features Maps in Deep Learning Architectures for Supporting Covid-19 Diagnosis C1
Iñaki Rodríguez-García, Tomás Sánchez-Pastor, Julia Vázquez-Escobar, Juan Luis Gómez-González, and Miguel Cárdenas-Montes

Uncertainty Propagation and Salient Features Maps in Deep Learning Architectures for Supporting Covid-19 Diagnosis

Iñaki Rodríguez-García, Tomás Sánchez-Pastor, Julia Vázquez-Escobar, Juan Luis Gómez-González, and Miguel Cárdenas-Montes

Abstract Doubt and the ability to point out details that identify an object or categories of objects are peculiarities of human intelligence. Roughly speaking, artificial intelligence aims to mimic the behavior of human intelligence. This work is a first attempt at joint use of previously existing technologies to mimic these characteristics of human intelligence. This work aims to help in the diagnosis of X-ray chest images with pneumonia, and covid-19; and images of healthy individuals by applying deep neural networks. These deep neural networks are modified so that they can generate predictions with uncertainty. Subsequently, on the previously generated predictions, the salient feature maps are generated to identify on which parts of the image the forecast decision is based. As a result of the work, examples of X-ray chest images will be shown where independent executions predict different labels, focusing attention on different areas of the radiography. So that the different areas indicated by the independent runs might help in the diagnosis of the different pathologies.

The original version of this chapter was revised: Author provided belated corrections have been incorporated. The correction to this chapter is available at https://doi.org/10.1007/978-3-031-04597-4_12

I. Rodríguez-García · T. Sánchez-Pastor · J. Vázquez-Escobar · J. L. Gómez-González · M. Cárdenas-Montes (✉)
Centro de Investigaciones Energéticas Medioambientales y Tecnológicas, Madrid, Spain
e-mail: miguel.cardenas@ciemat.es

I. Rodríguez-García
e-mail: inaki.rodriguez@ciemat.es

T. Sánchez-Pastor
e-mail: tomas.sanchez@ciemat.es

J. Vázquez-Escobar
e-mail: julia.vazquez@ciemat.es

J. L. Gómez-González
e-mail: juanluis.gomez@ciemat.es

V. Chang et al. (eds.), *Artificial Intelligence and Machine Learning Methods in COVID-19 and Related Health Diseases*, Studies in Computational Intelligence 1023,
https://doi.org/10.1007/978-3-031-04597-4_1

Keywords Deep learning · Decision making · Convolutional neural networks · Inception · Uncertainty propagation · Grad-CAM · Salient map · Covid-19

1 Introduction

Artificial Intelligence (AI) tries to mimic any of the aspects that we associate with human intelligence. High qualified human individuals spend much time doing repetitive but rather high expertise requirement tasks. These are extremely valuable and important actions, i.e. diagnosing potentially harmful diseases, managing complex systems as transport managements, resources distributions. Much training time is required for these tasks as in many cases high qualification requires years of study for just one individual. The role of AI to this regard as mimicking human intelligence, within the strong limitations of lacking creativity, opens the possibility for designing efficient decision making systems to automatically perform these repetitive tasks. This brings the possibility for individuals to discuss more creative problems or investing time in particular problems yet to be automatically solved. If there is an automatic way to invest resources qualified workers can invest time in innovative new ways to employ resources or to design an specific application for a particular problem. A clinician is more valuable investing time in their patients rather than evaluating chest x-rays in order to diagnose pneumonia or Covid19. In particular recent pandemics has proven that the capability of quickly diagnosing this illness fosters the favorable outcomes in a environment of scarce resources. Because of this in medicine AI is particularly valuable once an algorithm is trained as it can evaluate their decision within the matter of seconds and repetitively. AI represents a potentially real breakthrough in health systems for this sensible capability of decision making. Regardless of the well known ability to discern among possibilities in pre-trained models, uncertainty remains as a goal to tackle. In industry and research these techniques are ubiquitous, but there remains work and research into prescribing a proper measure of uncertainty in decision and prediction. Bringing back the medicine example, clinicians can determine that an X-ray image shows signs of pneumonia but simultaneously to express uncertainty regarding the decision, thus requiring other clinicians' expertise. A ML algorithm basically prescribes a recipe for a particular solution without discussion or in the best case providing robust uncertainty measure as a real probability (Alizadehsani et al., 2021; Abdar et al., 2021). Two aspects are highlighted in this work. On the one hand, the ability to synthetically express what elements a decision is based on: for example, the classification of an image under a certain label. And on the other hand, the ability to doubt our own decision and issue a second ranking in order of importance. Nowadays, artificial intelligence builds models that humbly aspire to address the ability to provide real probability altogether decisions.

In this work, started during the quarantine suffered in Spain during the first semester 2020 due to the covid-19 pandemic, the improvement of the diagnosis of covid-19 patients has been addressed. Starting from a set of X-ray chest images of patients diagnosed with pneumonia, covid-19 and healthy individuals, the construction of a model based on deep learning with the capacity to express uncertainty in predictions, as well as to highlight the relevant characteristics of an image so that it is classified under a certain label. In this case, it should be noted that the objective included making it possible to highlight the most relevant characteristics of the image when it is classified under the correct label, such as when an incorrect forecast is issued.

Although AI is being widely applied in science and industry, on rare occasions the predictions generated are accompanied by explanations based on the salient features directly impacting on the forecasting. However, these explanations are mandatory to reduce the impact of undesirable bias in deep learning based models (Murdoch et al, 2019). The presence of undetected bias leads to problems during the generalization of the model and therefore, in the industry adoption out of the research activity (Paleyes et al., 2020).

Another limitation encountered by deep learning algorithms is the computational cost required to include the intrinsic uncertainties of the data in their predictions. Most deep learning models are often viewed as deterministic functions, and as a result viewed as operating in a very different setting to the probabilistic models which possess uncertainty information. Despite the computational cost, Bayesian Neural Networks (BNN) provide probabilistic distributions of predictions, where uncertainties are taken into account. It is impossible to overemphasize the importance of this subject in medical imaging studies, where a confidence interval in diagnosis-classification scores could have a significant effect on the medical decision-making process.

Nevertheless, more efficient algorithms than BNN are gaining importance in the last years due to the rise of Stochastic Regularization Techniques (SRT). SRT uses dropout layers to smoothly break the deterministic outcomes of Neural Networks (Gal, 2016b). This approach overcomes one of the main barriers of BNN which is their computational intensity of humble problem sizes.

The outburst of the pandemic has caused many of the scientific efforts to be devoted to fight it. Accurate covid-19 detection is key to avoid contact with people that are infected and stop spreading it. Machine Learning has been widely used as a resource for this task (Kassani, 2020; Brinati et al., 2020; Gozes et al., 2020; Sánchez-Pastor et al., 2020). Specifically, convolutional neural networks make great classifiers for chest X-rays images (Panwar et al., 2020; Wang et al., 2020). Also CNNs has been used for covid-19 diagnosis from other image types such as Computed Tomography (CT) scans (He et al., 2020; Li et al., 2020).

The techniques used in this work to improve the results and clarify them have been used separately in other works. Inception nets are used for covid-19 detection in Das et al. (2020). Grad-CAM is also used for covid-19 detection in Basu et al. (2020) for the same purpose, checking that the important parts of the images for the net decision are coherent with respect to the opinions of experts. There is no use of the Bayesian approximation for covid-19 detection in the literature but there are some applications of the Bayesian Networks such as Butcher and Fenton (2020), where the probability of covid-19 infection is calculated and used for tracking the risk of infection in areas or triage of patients in hospitals.

Medical diagnosis is a sensitive application of machine learning that requires of explainability properties in its predictions. Robust uncertainty propagation is a desired property when performing predictions and classifications in medical diagnosis. The application of Bayesian Neural Networks has found a natural field of applicability in medical image classification and particular implementations are diabetic retinopathy detection (Wang and Yang, 2017) or ischemic stroke lesions (Kwon et al., 2018, 2020). Notwithstanding medical diagnosis spans beyond the realm of medical imaging. The robust propagation of uncertainty over the network makes this kind of neural network suitable for medical ontology probabilistic inference, as found for stroke diagnosis using ERM data in Shen et al. (2018) or liver disorders in Onisko et al. (1999). Ontology inference in the medical field also encompasses genetic semantics classification (Bauer et al., 2012).

The Spanish Strategy for Artificial Intelligence (ENIA), in the section IA for Society claims the use of Explainable Artificial Intelligence in any medicine related activity (Estrategia Nacional de Inteligencia Artificial, 2020).

> In this context, an AI capable of explaining to medical professionals the reasons for their decisions is necessary, improving human-computer interaction methods. Today, doctors draw on their experience to when making the most likely diagnosis. An automatic assistance, capable of carrying out said calculations of probability in a normative way and with real-time access to the necessary data from the scientific literature, it would allow greater productivity of health professionals. Training for the existence of a new generation of more technological doctors capable of helping in the design of these cognitive assistants is one of the challenges in this regard.
> The so-called "P4 Medicine" (predictive, personalized, preventive and participatory), will be based on technologies emerging markets such as AI and big data analytics based on machine learning and vision by computer. Thus, data science will be applied routinely to information, structured and unstructured, from electronic health records, -omics (genomics, proteomics, transcriptomics, etc.) and medical imaging.

In the present work, we apply SRT techniques to propagate the uncertainties in a CNN-Inception network in order to classify X-ray chest images labelled: pneumonia and covid-19 pathologies, and health individuals, at the same time that the most salient features of the images when classified under a certain label are highlighted. The rest of the work is organized as follows. In Sect. 2, the methods and the dataset used are explored. Firstly, the use of CNN-Inception, Uncertainty Propagation and Grad-CAM algorithms are explained and justified for this problem. Section 3 is devoted to discuss the obtained results for three different cases: pneumonia infected X-ray, covid-19 positive and healthy diagnosis X-ray. Finally, in Sect. 4 the conclusions are collected, as well as the future work.

2 Methods and Materials

2.1 CNN-Inception

Convolutional Neural Networks (CNNs) are inspired by the configuration of the animal visual cortex and are intended to treat images dataset (LeCun, 1989). Nowadays the classification of images is usually processed with this type of deep architecture. Its name comes from the fact that the hidden layers perform convolutions on the data.

The CNN consists of a sequence of convolutional layers, the output of which is connected only to local regions in the input. These layers alternate convolutional, non-linear and pooling-based layers which allow extracting the relevant features of the class of objects, independently of their placement in the data example. The CNN allows the model to learn filters that are able to recognize specific patterns in the input, and therefore they can capture richer information from the input. It also embodies three features which provide advantages over the multilayer perceptron: sparse interactions, parameter sharing and equivariance to translation (Goodfellow, 2016).

Although CNN are frequently associated to image or audio classification -2D grid examples- or video sequence -3D grid examples-, it can also be applied to time series analysis -1D grid examples-. When processing time series, instead of a set of images, the series has to be divided in overlapping contiguous time windows. These windows constitute the examples, where the CNN aims at finding patterns. At the same time, the application to time series modelling requires the application of 1D convolutional operators, whose weights are optimized during the training process.

Inception is a variant of CNN architecture frequently used in computer vision with a large number of successful cases (Szegedy, 2017; Szegedy et al., 2015). Inception architecture is composed in its main part by a set of repeated blocks. Each block is in turn composed of parallel branches of convolutional layers. Inside the Inception rationale, the branches are configured with different kernel sizes. The different sizes of the receptive fields in CNN aim to capture information at different scales: the larger the receptive field, the more generic are the features extracted from the images; whereas in the opposite sense, the smaller the kernel sizes, the more local are the features extracted.

To improve the performance of a CNN for image classification extending network depth and width can be considered, but this comes with disadvantages. By making the network more complex it takes the risk of overfitting. The final model may perform excessively well for the training data and not good enough when applied to test data. Furthermore, this increment in size will require more computational resources. If the resulting model has most of its weights almost null we are wasting the network resizing.

A larger fully connected neural network can be replaced by a smaller, sparsely connected one, efficiently constructed to reflect the data probability distribution. This is the problem of sparsity detection. It is treated in Arora et al. (2014) where the following result is presented: If the data can be represented by a very large and highly

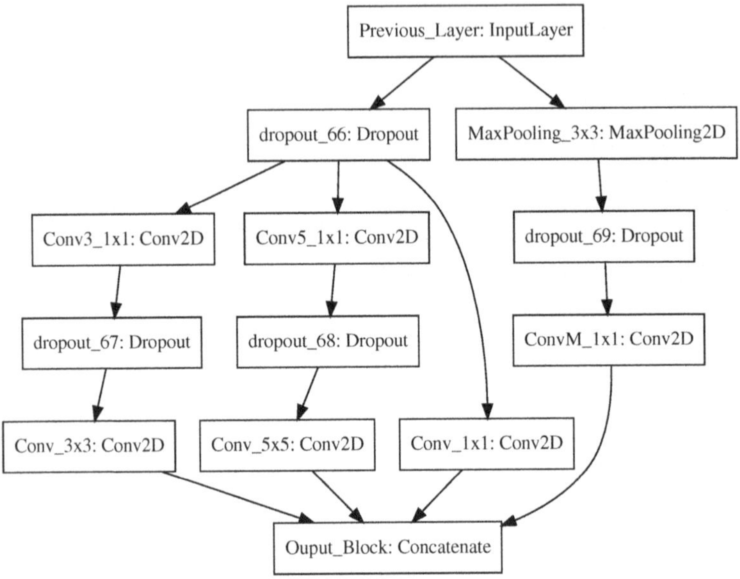

Fig. 1 Module Inception used in the proposal. Five Inception blocks compose the central part of the deep architecture

sparse neural network, then the optimal neural network can be built by studying the correlations of neurons in each layer. Nevertheless the numerical calculation of non-uniform sparse data compositions is very inefficient.

The Inception method for CNN architecture tries to exploit the data sparse configuration while maintaining the use of dense networks. It joins various dense networks with different configurations creating a non-uniform architecture.

In this way a big sparse network is approximated, that would have the mentioned drawbacks, with this design, more efficient and with great results. We would like to assess the correlation of the layers of the big sparse net and proceed using this information. The following construction is guided by these ideas and has great results for image classification. Traditional operations in CNNs are applied in parallel, forming a block substituting the conventional CNN layer (Fig. 1). Then these blocks are concatenated to form the complete algorithm (Fig. 2). This topology for the network is just an option from many others that could work as well or even better. This particular choice is merely based on the mentioned intuitions, so its success must be taken with caution.

In order to efficiently discriminate with the Inception module relevant features in the X-ray chest images at different size scales, filters of sizes 1×1, 3×3 and 5×5 are used in the branches of the module.

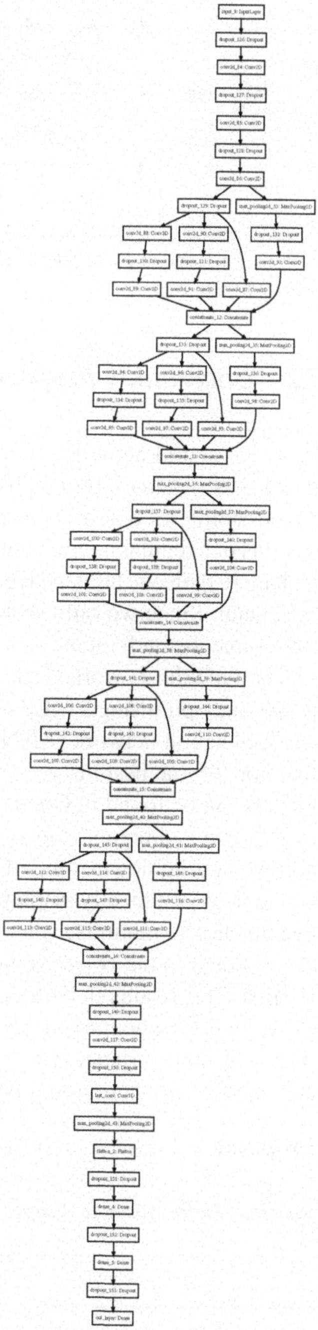

Fig. 2 Architecture used in the proposal. It can be appreciated the insertion of the Inception basis block five times. The stem section—just before the first Inception block—stacks three convolutional blocks composed of a convolutional layer and a dropout layer. After the five Inception blocks, two additional convolutional blocks identical to those used in the stem part, and two blocks of fully connected layers with dropout are inserted before the output layer. The output layer is composed of 3 neurons with softmax as activation function

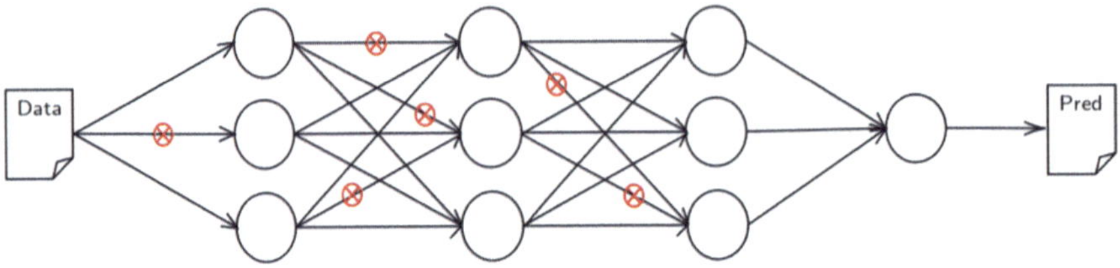

Fig. 3 Schema of dropout in a fully connected layer composed of tree hidden layers with three neurons per layer, and one single neuron in the output layer. The zeroed connection are pointed with a red signal

2.2 *Uncertainty Propagation*

Bayesian neural networks, first suggested in the '90s and then extensively studied (Neal, 1995), offer a probabilistic approach to deep learning models. They provide the possibility of obtaining uncertainty measures for the predictions by defining the weights of a regular neural networks as random variables. However, for architectures with more than one hidden layer, the extraction of these quantities become intractable in general. Since we only seek uncertainties of the final predictions, working with BNNs may be inefficient.

The PhD thesis work (Gal, 2016a) shows a correspondence between optimising BNNs and optimising a MLP with dropout applied. A MLP is a class of feedforward artificial neural network (ANN), a deep learning (DL) model whose task is to find a function f^* that best fits $y = f^*(x)$, given some data (x, y). More details about DL models can be found in Goodfellow (2016).

Dropout is a regularisation technique (Srivastava et al, 2014). It consists of randomly shutting down some of the connections between neurons when training the network, as graphically displayed in Fig. 3. It is proven in Gal (2016a) that the objective function to minimise when training a BNN can be approximated to an expression proportional to the objective function related to the training of a MLP with dropout applied. This results in both optimisation processes being equivalent.

Applying dropout to an ANN offers the possibility of obtaining expected predictions with their variance (Gal, 2016a). Two results of Gal (2016a) allow us to estimate the expected prediction of a BNN and its variance, given some training data.

Proposition 1 *Given* $p\left(\mathbf{y}^*|\mathbf{f}^{\omega}\left(\mathbf{x}^*\right)\right) = \mathcal{N}\left(\mathbf{y}^*; \mathbf{f}^{\omega}\left(\mathbf{x}^*\right), \tau^{-1}\mathbf{I}\right)$[1] *for some* $\tau > 0$, $\mathbb{E}_{q_{\theta}^*(\mathbf{y}^*|\mathbf{x}^*)}\left[\mathbf{y}^*\right]$ *can be estimated, if it is trained the MLP with dropout and make predictions for some test input* $\mathbf{x}^*$ T *times, with the unbiased estimator*

[1] Where $p\left(\mathbf{y}^*|\mathbf{f}^{\omega}\left(\mathbf{x}^*\right)\right)$ is the posterior probability distribution of the theoretical BNN prediction $\mathbf{y}^*$ conditioned to the output of the model $\mathbf{f}^{\omega}\left(\mathbf{x}^*\right)$.

It is supposed that it follows a normal distribution with mean $\mathbf{f}^{\omega}\left(\mathbf{x}^*\right)$ and variances $\tau^{-1}\mathbf{I}$, where $\mathbf{I}$ is the identity matrix.

$$\widetilde{\mathbb{E}}\left[\mathbf{y}^*\right] := \frac{1}{T}\sum_{t=1}^{T}\mathbf{f}^{\hat{\omega}_t}\left(\mathbf{x}^*\right) \underset{T\to\infty}{\longrightarrow} \mathbb{E}_{q_\theta^*(\mathbf{y}^*|\mathbf{x}^*)}\left[\mathbf{y}^*\right] \tag{1}$$

where $\mathbf{f}^{\hat{\omega}_t}(\mathbf{x}^*)$ *is the prediction of the dropout network for some new data* $\mathbf{x}^*$, *being* $\hat{\omega}_t$ *the optimal weights in each training* t, *and* $q_\theta^*(\mathbf{y}^*|\mathbf{x}^*)$ *is the optimal predictive distribution for the BNN.*

The second moment can also be estimated.

Proposition 2 *Following the notation from the previous proposition, given* $p(\mathbf{y}^*|\mathbf{f}^{\omega}(\mathbf{x}^*)) = \mathcal{N}\left(\mathbf{y}^*; \mathbf{f}^{\omega}(\mathbf{x}^*), \tau^{-1}\mathbf{I}\right)$ *for some* $\tau > 0$, $\mathbb{E}_{q_\theta^*(\mathbf{y}^*|\mathbf{x}^*)}\left[(\mathbf{y}^*)^T(\mathbf{y}^*)\right]$ *can be estimated, proceeding the same as before, with the unbiased estimator*

$$\widetilde{\mathbb{E}}\left[(\mathbf{y}^*)^T(\mathbf{y}^*)\right] := \tau^{-1}\mathbf{I} + \frac{1}{T}\sum_{t=1}^{T}\mathbf{f}^{\hat{\omega}_t}(\mathbf{x}^*)^T\mathbf{f}^{\hat{\omega}_t}(\mathbf{x}^*) \underset{T\to\infty}{\longrightarrow} \mathbb{E}_{q_\theta^*(\mathbf{y}^*|\mathbf{x}^*)}\left[(\mathbf{y}^*)^T(\mathbf{y}^*)\right] \tag{2}$$

Having the second moment, the variance of the prediction can be obtained as:

$$\widetilde{\mathrm{Var}}\left[\mathbf{y}^*\right] := \tau^{-1}\mathbf{I} + \frac{1}{T}\sum_{t=1}^{T}\mathbf{f}^{\hat{\omega}_t}(\mathbf{x}^*)^T\mathbf{f}^{\hat{\omega}_t}(\mathbf{x}^*) - \widetilde{\mathbb{E}}\left[\mathbf{y}^*\right]^T\widetilde{\mathbb{E}}\left[\mathbf{y}^*\right] \underset{T\to\infty}{\to} \mathrm{Var}_{q_\theta^*(\mathbf{y}^*|\mathbf{x}^*)}\left[\mathbf{y}^*\right] \tag{3}$$

In the cited work (Gal, 2016a) the term τ is calculated with a formula containing the parameters selected for the network. In Zhu and Zhu (2022) this quantity is estimated by residual sum of squares, evaluated in a separate validation set with size V (Eq. 4).

$$\tau = \frac{1}{V}\sum_{v=1}^{V}\left(\mathbf{y}_v^* - \mathbf{f}^{\hat{\omega}_t}\left(\mathbf{x}_v^*\right)\right)^2 \tag{4}$$

This way, the use of the parameters of the network that depend on their prior distribution is avoided, this frequentist approach is chosen.

This means that only needs to train a regular MLP with dropout applied various times and then estimates the mean and the variance of the theoretical BNN.

In the past, this methodology has been applied to classification problems in High Energy Physics in the context of CMS collaboration (Vázquez-Escobar et al., 2021).

2.3 *Grad-CAM*

In this work a complex model that has a great performance and also gives results with uncertainty has been built. Now the objective is to understand how the network classifies the data, check which features are relevant and if it makes sense. By constructing a transparent model accidental or biased success can be identified. We resort to the Grad-CAM method (Selvaraju et al., 2019) to achieve this. It provides explanations for CNN models, in this case for image classification. For every image it computes a gradient based on the importance of its pixels, resulting in a visual and intuitive map, easy to interpret. It is usually represented as a heatmap over the image.

CNNs are formed by convolutional layers that, unlike fully connected layers, collect spatial information. CNNs process local features from images to capture high-level knowledge about them. Grad-CAM usually uses the last convolutional layer of the network, computes the gradient of the score for a class with respect to feature map activations, then these gradients flowing back are global-average-pooled over the layer dimensions to obtain the importance of the neurons for the decision regarding this class. This is reflected in Eq. 5 where the neuron importance weights α_k^c is calculated, y^c is the score for the class c and A^k are the map activations.

$$\alpha_k^c = \frac{1}{Z} \sum_i \sum_j \frac{\partial y^c}{\partial A_{ij}^k} \tag{5}$$

Since α_k^c corresponds only to the feature map A^k, it only represents its importance for the class c. For complete information for the class c, a linear combination of the weights is computed (Eq. 6).

$$L_{\text{Grad-CAM}}^c = ReLU\left(\sum_k \alpha_k^c A^k\right) \tag{6}$$

This results in the mentioned heatmap over the original image, the object $L_{\text{Grad-CAM}}^c$ has the same size as the activations A^k. A $ReLU$ function is applied because only positive terms are considered to be influential. Every pixel of the original image and the part it plays in the decision of the CNN for the class c can be studied.

3 Experimental Results and Models Comparison

In the following examples, the salient maps or the regions of the images that contribute to the classification of the image as a certain true label are depicted and discussed. Examples for the three labels handled: pneumonia, covid-19 and healthy are shown. Although in most of the cases the X-ray images are correctly classified, a case of misclassification is also discussed. We must notice the discussion about this case: a pneumonia label predicted as covid-19 for its enriching role.

The final architecture used in the proposal is composed of a stem section, five Inception blocks (Fig. 1), a section before the output layer and the output layer (Fig. 2). The stem section stacks three convolutional blocks composed of a convolutional layer and a dropout layer. After the five Inception blocks, two additional

Table 1 Number of images of each label in training and test sets before balancing the minority class

Dataset	Healthy	Pneumonia	Covid-19
Training	7966	5448	187
Test	885	592	64

convolutional blocks identical to those used in the stem part, and two blocks of fully connected layers with dropout are inserted before the output layer. The output layer is composed of 3 neurons with `softmax` as activation function, each one handles one of the labels. Labels are one-hot encoding. Diverse variants of this architecture were exhaustively tested. Deeper and wider network architectures critically increased the processing time without an appreciated improvement in the performance, where as the shallower ones degraded their performance. Therefore, the architecture selected is a trade-off between performance and processing time in the computational resources available.

X-ray images have been recollected from diverse repositories (North America, 2019; Chung et al., 2020; Cohen et al., 2020). Regarding the dataset labels, it is strongly imbalanced with the label covid-19 as minority class. The training set contains 7966 without pathologies, 5447 pneumonia and 187 covid-19 positive (Table 1). Due to the big imbalance of the no-covid versus covid-19 images, an oversampling method has been applied.

The original images are rescaled to a size of 200×200 pixels for standardization purpose. To fix the imbalance, new images of the minority class are created by zooming the original ones. Rectangles in the images of size 180×180 pixels with a step of 5 pixels are created. This allows creating 25 new images per minority class instance (Fig. 4). Thus, the number of images of the minority class can be balanced with the number of images of the other classes. Also, new images of 180×180 pixels of the majority classes are created by randomly zoomed around the central area, but in this case a single image is created. For the images of the majority classes, these new images are used in the classification, while the original ones of 200×200 pixels are discarded. Later a randomly-selected balanced set of images is used for the training and prediction with the deep architecture. This tries to avoid undesirable bias due to the zooming process it-self.

In addition to the analysis of the salient features of selected cases, an numerical analysis of the scores of the predictions with their associated uncertainty is presented. This analysis includes cases of straightforward predictions and confusing ones.

3.1 *Pneumonia Image*

In Fig. 5, a first example of the application of the joint of Grad-CAM technique and CNN with Inception modules an uncertainty propagation is shown. The X-ray chest

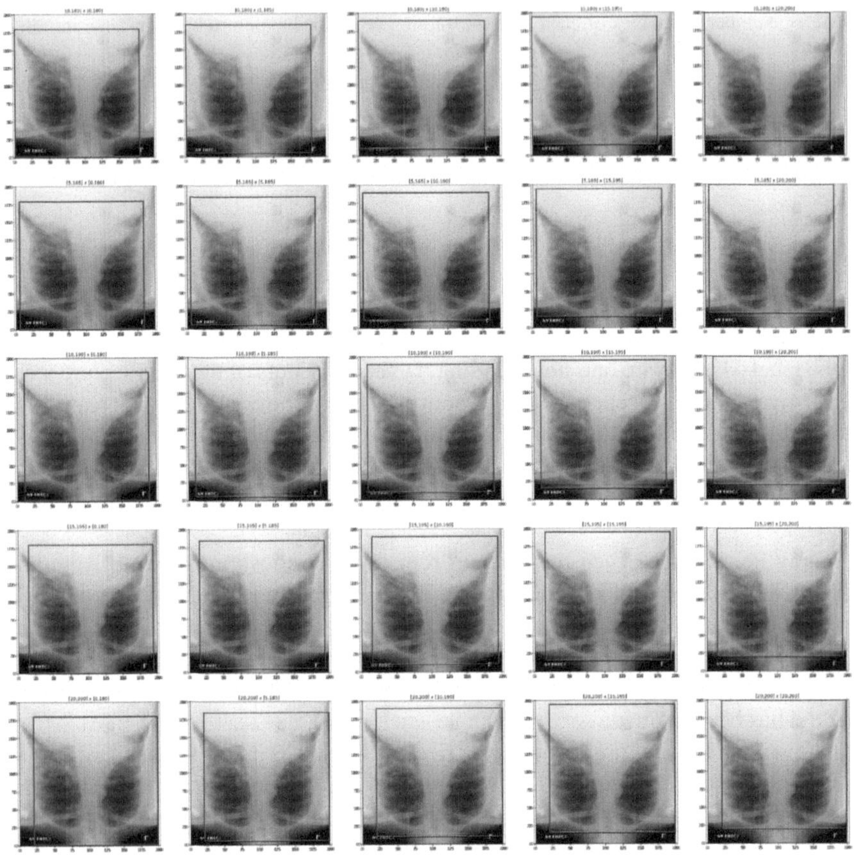

Fig. 4 Example of 25 images created by zooming. An area of size of 180 × 180 pixels is selected, moved with steps of 5 pixels around the center of the image of original size 200 × 200 pixels

image is presented in the top-left panel, whereas the salient maps emerging from the prediction of five independent runs are presented in the other panels. Four of the five runs predict the true label—pneumonia—with probabilities from 0.51 to 0.71. A single run incorrectly predicts the label covid-19 with a probability of 0.511 (bottom-central panel), at the same time that in this run the probability of pneumonia goes down to 0.486.

For the correctly predicted labels (top- central and right panels, and bottom- left and right panels), the salient maps highlight a bottom area of the right lung, and a smaller area in the inferior part of the left lung. With regard to the case incorrectly classified as covid-19 (bottom-central panel), a completely different area is illuminated in the middle part of the left lung.

This example illustrates the focus of this work: to generate predictions with some doubt and with the ability to indicate which part of the input data is most sensitive to

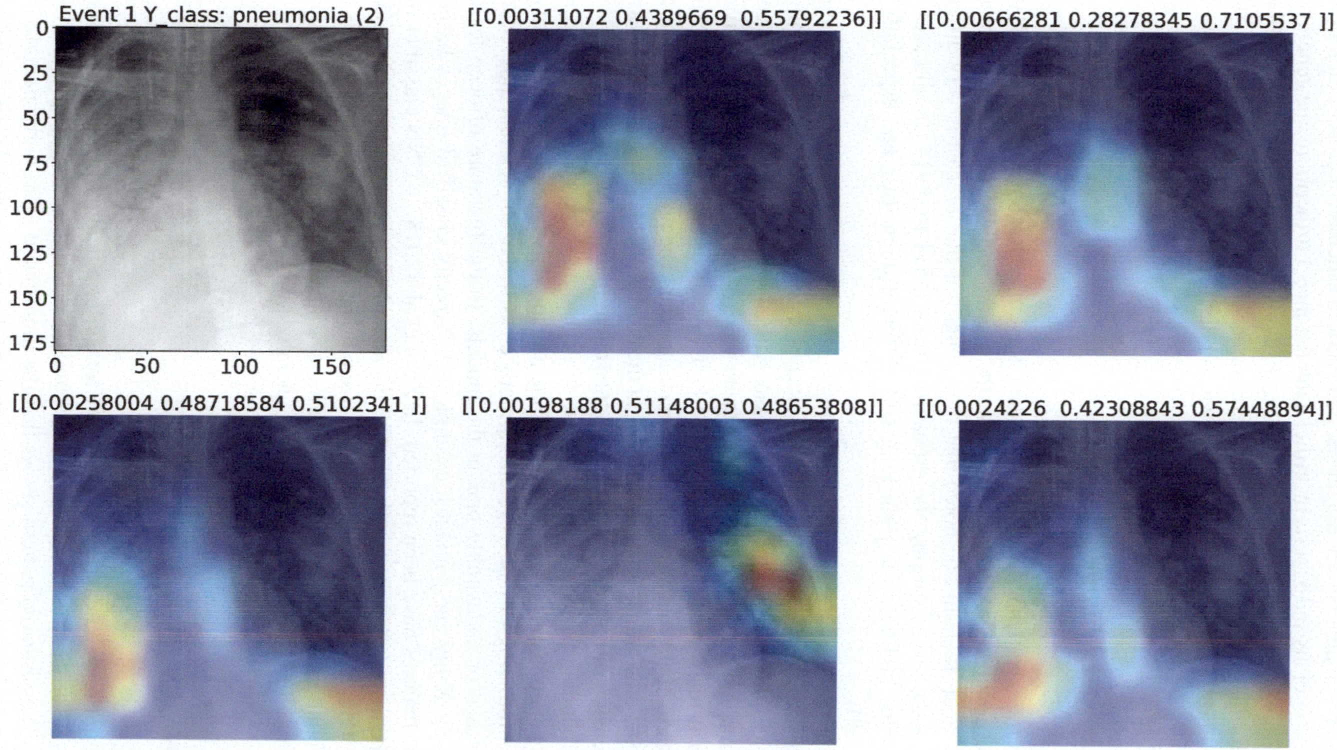

Fig. 5 Example of application of the proposed methodology to X-ray chest corresponding to the label pneumonia (top-left panel), and its predictions produced by five independent runs. Four runs correctly predict the label (top- central and right panels, and bottom- left and right panels). For those predictions the salient feature maps exhibit a high similarity. For one run an incorrect prediction is cast (bottom-central panel) for which the predicted label is covid-19. In this case, the salient map for this label is disjoint with the areas of the pneumonia label

the forecast for each predicted class: true and false. We consider that this information is very valuable for the diagnosis, and it could articulate a closer collaboration between clinicians and experts in artificial intelligence. Furthermore, the general proposal of this work can be exported to other diseases involving data of very different nature, for instance genomics and transcriptomics. In the following examples, other cases with different true classes are shown and discussed.

3.2 *Covid-19 Image*

In Fig. 6, the salient maps of five independent runs corresponding to a X-ray chest with true label covid-19 are shown. In this case, the five independent runs correctly predict the true label. It should be noted that the strength of the forecast is very different between executions, varying the probability from 0.511 to 0.914. In all cases, the salient maps highlight the right lung with different intensities, and especially the bottom part.

As it can be appreciated, the predictions achieving high probability of the true label, for instance the left-bottom panel with a probability of 0.914, tend to illuminate the whole right lung. Conversely, in the cases with a lower probability, for instance the central-top and the right-bottom panels (with score of 0.578 and 0.511 respectively), the salient maps concentrate around the bottom part of the right lung, and with a lower intensity around the top of the same lung. Thus, the proposed approach is able to translate the grading in the probability of the most probable label to intensity levels and extension of the area illuminated in the salient maps.

Confounding variables can contribute to degrading the performance when generalizing deep learning-based models. Salient maps visualization allows identifying these confounding variables even in correctly classified examples, and thus reducing the presence of undesirable bias in the models. In the literature of deep learning classifying X-ray chest images, examples of correctly classified images based on spurious features have been reported. In Zech et al. (2018), a CNN classifier highlights the metal token that radiology technicians place on the patient in the top left corner of the image as a salient feature. It should be noticed the absence of this type of confounding variables at least in the images inspected by the authors. This is a necessary check for ensuring a correct generalization of the model.

3.3 *Healthy Image*

Finally in Fig. 7, an example of an X-ray image for a healthy person. The question for this label is, which area might be illuminated in the salient map of a healthy person. The observed pattern shows that the illuminated areas correspond to those areas disjointly out of the organs illuminated of the patients with labels associated to the illnesses: covid-19 and pneumonia. Figure 7 illustrates this observation: the five

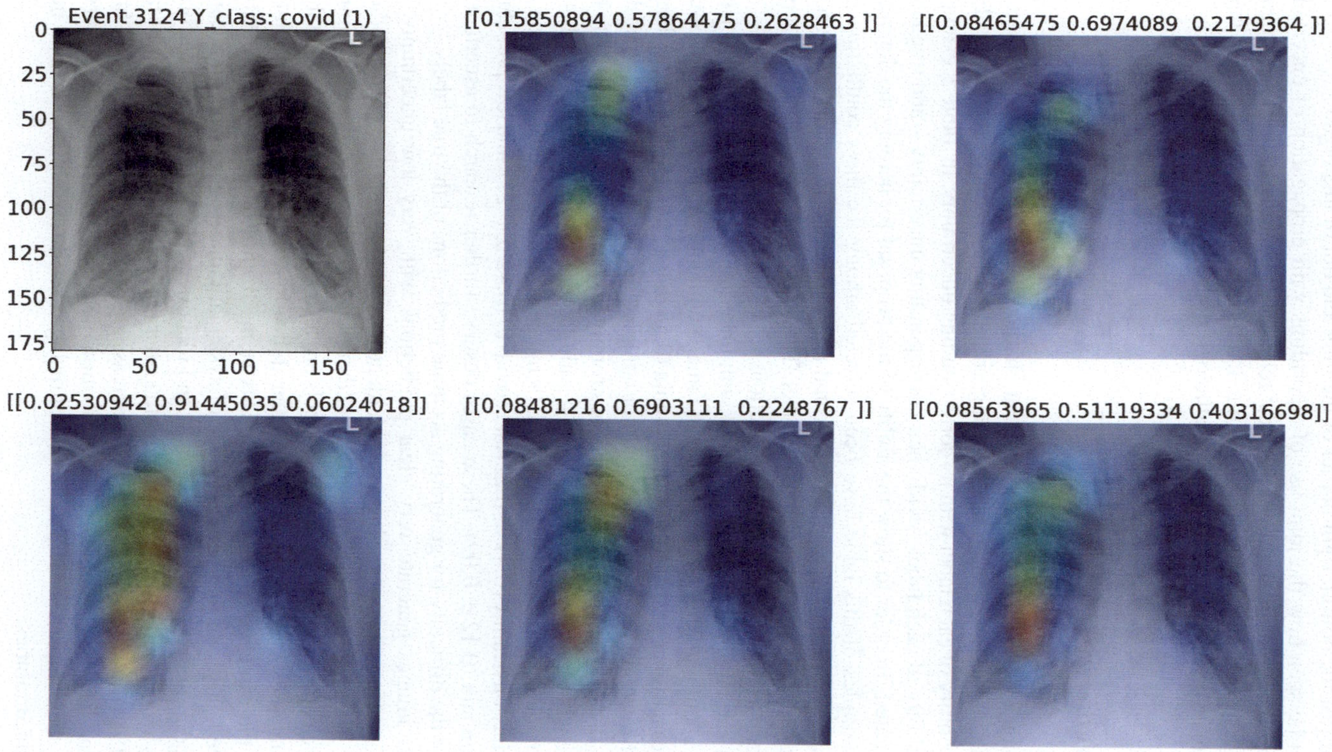

Fig. 6 Example of application of the proposed methodology to X-ray chest corresponding to the label covid-19 (top-left panel), and its predictions produced by five independent runs. In this case the five runs correctly predict the true label. The salient feature maps exhibit a high similarity, illuminating repetitively the bottom part of the right lung

independent runs correctly predict the true label (healthy), while the salient maps highlight just the bottom right border of the lung and out of this organ.

It can be seen how the joint use of algorithms with propagation of uncertainty altogether the ability to highlight the most relevant characteristics for the particular forecast issued allows us to understand the doubt between different forecasts based on the areas afflicted with the various pathologies or labels.

3.4 Numerical Outputs

In addition to display the most prominent characteristics in each X-ray image for its prediction under a certain label, the algorithm issues a numerical forecast about the probability of each label with its corresponding uncertainty. In Table 2, examples of probability of each label—with their corresponding uncertainty—and true labels are presented. In order to show the ability of the proposal for helping the diagnosis, some examples of this table have been selected among the most conflicting predictions. Other examples have been selected under the opposite criterion: the score predicts with high probability the true label. They are separately shown. The scores are generated from 8 independent predictions.

As it can be appreciated, predictions with uncertainty can be used for decision-making about pathology. For instance, the image with identifier 970—pneumonia—expresses a higher probability of pneumonia (0.58 ± 0.17) than for the other labels. Being the second most probable, and not completely disposable, the label healthy (0.41 ± 0.18); whereas covid-19 label is strongly discarded (0.01 ± 0.06). Another example with strong confusion between a pathological label and healthy label is the image of identifier 1489. This is a covid-19 image with almost equal prediction and uncertainty on healthy (0.47 ± 0.22) and covid-19 (0.42 ± 0.27), and low probability in pneumonia (0.12 ± 0.09). These cases deserve further research to ascertain the origin of the difficulty for generating accurate predictions of the true labels.

Other images generate very clear predictions with very low uncertainty. For instance, the predictions for images with identifiers 94, 408, and 583 clearly indicate healthy label, which is the true label, and therefore non illness evidence. Examples of very clear predictions of the other labels are also shown in this part of the table.

In the cases of predictions with a clear indication of a class, the uncertainty is almost the same for the three labels. This stems from a null presence of statistical uncertainty, and the mere presence of systematic uncertainty. The same can be observed in the cases of label covid-19 and identifiers 2753 and 1908. The former one has a score of 0.99 ± 0.12, whereas the latter one reduces the probability of label covid-19 at the same time that slightly increases the uncertainty: 0.97 ± 0.13. In the latter one, a small but appreciable contribution of statistical uncertainty is observed. Finally the comparison with the most confusing cases throws out scores with much larger uncertainties.

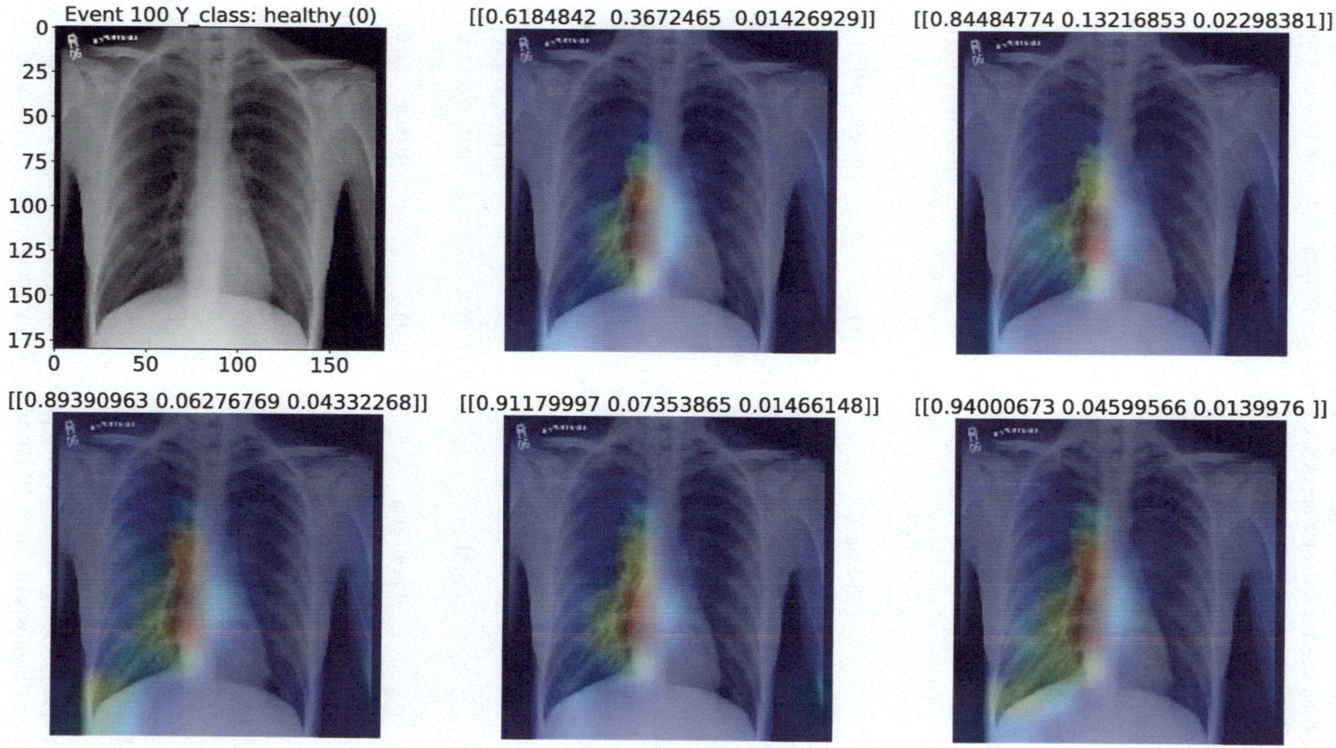

Fig. 7 Example of application of the proposed methodology to X-ray chest corresponding to the label healthy patient (top-left panel), and its predictions produced by five independent runs. In this case the five runs correctly predict the true label. The salient feature maps exhibit a high similarity, illuminating repetitively an area not completely covering a small part of the right lung

Table 2 Examples of forecasting healthy, pneumonia and covid-19 labels: label probability with uncertainty. Some examples have been selected among the most conflicting predictions, whereas other have been selected under the opposite criterion: the score predicts with high probability the true label. Results obtained from 8 independent predictions

Image id.	True label	One-hot coded	Prediction and uncertainty
Confusing cases			
1489	Covid-19	(0, 1, 0)	$(0.47 \pm 0.22, 0.42 \pm 0.27, 0.12 \pm 0.09)$
1519	Covid-19	(0, 1, 0)	$(0.13 \pm 0.12, 0.32 \pm 0.21, 0.55 \pm 0.20)$
1483	Covid-19	(0, 1, 0)	$(0.30 \pm 0.20, 0.59 \pm 0.26, 0.11 \pm 0.10)$
970	Pneumonia	(0, 0, 1)	$(0.41 \pm 0.18, 0.01 \pm 0.06, 0.58 \pm 0.17)$
1108	Pneumonia	(0, 0, 1)	$(0.46 \pm 0.10, 0.36 \pm 0.12, 0.19 \pm 0.09)$
942	Pneumonia	(0, 0, 1)	$(0.03 \pm 0.07, 0.64 \pm 0.31, 0.33 \pm 0.32)$
Clear cases			
94	Healthy	(1, 0, 0)	$(0.94 \pm 0.13, 0.05 \pm 0.13, 0.01 \pm 0.12)$
408	Healthy	(1, 0, 0)	$(0.99 \pm 0.12, 0.00 \pm 0.12, 0.01 \pm 0.12)$
583	Healthy	(1, 0, 0)	$(0.99 \pm 0.12, 0.00 \pm 0.12, 0.01 \pm 0.12)$
2106	Covid-19	(0, 1, 0)	$(0.00 \pm 0.12, 1.00 \pm 0.12, 0.00 \pm 0.12)$
2753	Covid-19	(0, 1, 0)	$(0.01 \pm 0.12, 0.99 \pm 0.12, 0.00 \pm 0.12)$
1908	Covid-19	(0, 1, 0)	$(0.02 \pm 0.13, 0.97 \pm 0.13, 0.01 \pm 0.12)$
1038	Pneumonia	(0, 0, 1)	$(0.00 \pm 0.12, 0.01 \pm 0.10, 0.99 \pm 0.12)$
1437	Pneumonia	(0, 0, 1)	$(0.01 \pm 0.12, 0.00 \pm 0.10, 0.99 \pm 0.12)$
1290	Pneumonia	(0, 0, 1)	$(0.00 \pm 0.12, 0.00 \pm 0.10, 1.00 \pm 0.12)$

4 Conclusions

In this paper, an effort to mimic two peculiarities of the human intelligence while helping the diagnosis of covid-19 based on X-ray chest image is presented. For this purpose, diverse techniques of artificial intelligence have been blended. Firstly the capacity of generation of prediction with uncertainty. And then the ability to highlight which part of each image is most relevant to that forecast. As a result of the work, artificial intelligence experts can assist clinicians in the diagnosis of covid-19.

To demonstrate the capability of the proposal, the results of three examples corresponding to the three labels handled—pneumonia, covid-19 and health—-have been shown. It can be seen how the joint use of algorithms with propagation of uncertainty and the ability to highlight the most relevant characteristics for the particular forecast issued allow us to understand the doubt between different predictions based on the areas with the various pathologies or labels. The selected cases correspond with some of the most confusing ones.

For other cases, only the numerical results associated with the probability—and the associated uncertainty—for each label are shown. For these cases, images with straightforward and confusing predictions have been underlined. The overall results demonstrate the ability of the proposal for generating accurate predictions at the same time that helping the clinicians in the diagnosis.

Additional efforts for implementing more efficient deep learning techniques are being carried out in the pathomics area, mainly in oncology. Broadening from image, as input for the deep architecture, to genomics and transcriptomics are being used to generate predictions with strong support on their explainability.

Acknowledgements IRG is funded through the PEJ2018-003089-P project to promote Young Employment and Implementation of the Youth Guarantee in R+D+i within the framework of the State Subprogram for the Incorporation of the State Program for the Promotion of Talent and Employability in R+D+i within the framework of the State Plan for Scientific and Technical Research and Innovation 2017–2020. This contract is co-funded by the European Social Fund and the Youth Employment Initiative through the Youth Employment Operational Program. TSP is co-funded in a 91.89% by the European Social Fund within the Youth Employment Operating Program, as well as the Youth Employment Initiative (YEI), and co-found in a 8.11 by the "Comunidad de Madrid (Regional Government of Madrid)" through the project PEJ-2018-AI/TIC-10290. JLGG is co-funded in a 91.89% by the European Social Fund within the Youth Employment Operating Program, as well as the Youth Employment Initiative (YEI), and co-found in a 8,11 by the "Comunidad de Madrid (Regional Government of Madrid)" through the project PEJ-2019-AI/TIC-12440. JVE and MCM are funded by the Spanish Ministry of Economy and Competitiveness (MINECO) for funding support through the grant "Unidad de Excelencia María de Maeztu": CIEMAT - FÍSICA DE PARTÍCULAS through the grant MDM-2015-0509, and the Spanish Ministry of Science and Innovation for funding support through the grant PID2020-113807RA-I00 "SERVICIOS INNOVADORES DE ANALISIS DE DATOS PARA EL EXPERIMENTO CMS".

References

Abdar, M., Pourpanah, F., Hussain, S., Rezazadegan, D., Liu, L., Ghavamzadeh, M., & Nahavandi, S. (2021). A review of uncertainty quantification in deep learning: Techniques, applications and challenges. *Information Fusion, 76*, 243–297. Retrieved from https://doi.org/10.1016/j.inffus.2021.05.008

Alizadehsani, R., Roshanzamir, M., Hussain, S., Khosravi, A., Koohestani, A., Zangooei, M. H., & Acharya, U. R. (2021). Handling of uncertainty in medical data using machine learning and probability theory techniques: A review of 30 years (1991–2020). *Annals of Operations Research*. Retrieved from https://doi.org/10.1007/s10479-021-04006-2

Arora, S., Bhaskara, A., Ge, R., & Ma, T. (2014, 22-24 Jun). Provable bounds for learning some deep representations. In E. P. Xing & T. Jebara (Eds.), *Proceedings References 19 of the 31st international conference on machine learning* (Vol. 32, pp. 584-592). Bejing, China: PMLR.

Basu, S., Mitra, S., & Saha, N. (2020). Deep learning for screening covid-19 using chest x-ray images. In *2020 IEEE symposium series on computational intelligence* (ssci) (pp. 2521–2527).

Bauer, S., Kohler, S., Schulz, M. H., & Robinson, P. N. (2012). Bayesian ontology querying for accurate and noise-tolerant semantic searches. *Bioinformatics, 28*(19), 2502–2508. Retrieved from https://doi.org/10.1093/bioinformatics/bts471

Brinati, D., Campagner, A., Ferrari, D., Locatelli, M., Banfi, G., & Cabitza, F. (2020). Detection of covid-19 infection from routine blood exams with machine learning: A feasibility study. *Journal of Medical Systems, 44*(8), 1–12.

Butcher, R., & Fenton, N. (2020). Extending the range of symptoms in a bayesian network for the predictive diagnosis of covid-19. *medRxiv*.

Chung, A., Wang, L., Wong, A., Lin, Z. Q., McInnis, P., & Gunraj, H. (2020). *Figure 1 Covid-19 chest x-ray data initiative*. Retrieved from https://github.com/agchung/Figure1-COVID-chestxray-dataset

Cohen, J. P.,Morrison, P., & Dao, L. (2020). *Covid-19 image data collection*. Retrieved from https://github.com/ieee8023/covid-chestxray-dataset

Das, D., Santosh, K., & Pal, U. (2020). Truncated inception net: Covid-19 outbreak screening using chest x-rays. *Physical and Engineering Sciences in Medicine, 43*(3), 915–925.

Estrategia Nacional de Inteligencia Artificial (Tech. Rep.). (2020, Nov.). Secretaría General Técnica del Ministerio de Ciencia, Innovación y Universidades.: Gobierno de España.

Gal, Y. (2016a). Bayesian deep learning. In *Uncertainty in deep learning* (pp. 29–61).

Gal, Y. (2016b). *Uncertainty in deep learning* (Unpublished doctoral dissertation). Cambridge: University of Cambridge.

Goodfellow, I. J., Bengio, Y., & Courville, A. (2016). *Deep learning*. Cambridge, MA, USA: MIT Press. http://www.deeplearningbook.org

Gozes, O., Frid-Adar, M., Sagie, N., Zhang, H., Ji, W., & Greenspan, H. (2020). Coronavirus detection and analysis on chest ct with deep learning. arXiv:2004.02640.

He, X., Yang, X., Zhang, S., Zhao, J., Zhang, Y., Xing, E., & Xie, P. (2020). Sampleefficient deep learning for covid-19 diagnosis based on ct scans. *medRxiv*.

Hinton, G. E., & van Camp, D. (1993). Keeping the neural networks simple by minimizing the description length of the weights. In *Proceedings of the sixth annual conference on computational learning theory* (pp. 5–13). New York, NY, USA: Association for Computing Machinery, https://doi.org/10.1145/168304.168306.

Kassani, S. H., Kassasni, P. H., Wesolowski, M. J., Schneider, K. A., & Deters, R. (2020). *Automatic detection of coronavirus disease (covid-19) in x-ray and ct images: A machine learning-based approach*.

Kullback, S. (1959). *Information theory and statistics*. New York: Wiley

Kwon, Y., Won, J.-H., Kim, B. J., & Paik, M. C. (2018). *Uncertainty quantification using bayesian neural networks in classification: Application to ischemic stroke lesion segmentation*.

Kwon, Y., Won, J.-H., Kim, B. J., & Paik, M. C. (2020). Uncertainty quantification using bayesian neural networks in classification: Application to biomedical image segmentation. *Computational Statistics and Data Analysis, 142*, 106816. https://doi.org/10.1016/j.csda.2019.106816.

LeCun, Y. (1989). *Generalization and network design strategies* (Tech. Rep. No. CRGTR- 89-4). Department of Computer Science, University of Toronto, 10 Kings College Rad, Toronto, Canada: University of Toronto.

Li, L., Qin, L., Xu, Z., Yin, Y., Wang, X., Kong, B., & Xia, J. (2020). Using artificial intelligence to detect covid-19 and community-acquired pneumonia based on pulmonary ct: Evaluation of the diagnostic accuracy.*Radiology, 296*(2), E65–E71. (PMID: 32191588). https://doi.org/10.1148/radiol.2020200905.

Murdoch, W. J., Singh, C., Kumbier, K., Abbasi-Asl, R., & Yu, B. (2019). Definitions, methods, and applications in interpretable machine learning. *Proceedings of the National Academy of Sciences, 116*(44), 22071–22080. https://doi.org/10.1073/pnas.1900654116

Neal, R. M. (1995). *Bayesian learning for neural networks* (Unpublished doctoral dissertation). University of Toronto.

North America, R. S. (2019). *Rsna pneumo- nia detection challenge.* Retrieved from https://www.kaggle.com/c/rsna-pneumonia-detection-challenge/data

Onisko, A., Druzdzel, M. J., D, P., Wasyluk, H., & D, P. (1999). *A bayesian network model for diagnosis of liver disorders* (Tech. Rep. No. CBMI-99-27). University of Pittsburgh, Pittsburgh, PA 15260, U.S.A: Center for Biomedical Informatics, University of Pittsburgh.

Paleyes, A., Urma, R., & Lawrence, N. D. (2020). Challenges in deploying machine learning: a survey of case studies. *CoRR*. ArXiv:abs/2011.09926. Retrieved from https://arxiv.org/abs/2011.09926.

Panwar, H., Gupta, P., Siddiqui, M. K., Morales-Menendez, R., & Singh, V. (2020). Application of deep learning for fast detection of covid-19 in x-rays using ncovnet. *Chaos, Solitons & Fractals, 138*, 109944.

Robbins, H., & Monro, S. (1951). A stochastic approximation method. *Annals of Mathematical Statistics, 22*, 400–407.

Rubin, D. B. (1981, 01). The bayesian bootstrap. *Annals of Statistics, 9*(1), 130–134. Retrieved from https://doi.org/10.1214/aos/1176345338

Rumelhart, D. E., Hinton, G. E., & Williams, R. J. (1986). Learning Representations by Back-propagating Errors. *Nature, 323*(6088), 533–536. Retrieved from http://www.nature.com/articles/323533a0, https://doi.org/10.1038/323533a0.

Sánchez-Pastor, T., Rodríguez-García, I., & Cárdenas-Montes, M. (2020). Clasificación de radiografías covid con inteligencia artificial. *Vértices, 33*, 24–26.

Selvaraju, R. R., Cogswell, M., Das, A., Vedantam, R., Parikh, D., & Batra, D. (2019). Grad-cam: Visual explanations from deep networks via gradientReferences 21 based localization. *International Journal of Computer Vision,* 128(2), 336- 359. Retrieved from http://dx.doi.org/10.1007/s11263-019-01228-7, https://doi.org/10.1007/s11263-019-01228-7.

Shen, Y., Zhang, L., Zhang, J., Yang, M., Tang, B., Li, Y., & Lei, K. (2018). Cbn: Constructing a clinical bayesian network based on data from the electronic medical record. *Journal of Biomedical Informatics, 88*, 1–10. https://doi.org/10.1016/j.jbi.2018.10.007.

Srivastava, N., Hinton, G., Krizhevsky, A., Sutskever, I., & Salakhutdinov, R. (2014). Dropout: A simple way to prevent neural networks from overfitting. *Journal of Machine Learning Research, 15*, 1929–1958.

Szegedy, C., Ioffe, S., Vanhoucke, V., & Alemi, A. (2017). Inception-v4, inception-resnet and the impact of residual connections on learning. In *Proceedings of the AAAI conference on artificial intelligence* (Vol. 31(1)).

Szegedy, C., Liu, W., Jia, Y., Sermanet, P., Reed, S., Anguelov, D., & Rabinovich, A. (2015). Going deeper with convolutions. In *Computer vision and pattern recognition (cvpr)*. Retrieved from http://arxiv.org/abs/1409.4842.

Tishby, L., & Solla. (1989). Consistent inference of probabilities in layered networks: predictions and generalizations. In *International 1989 joint conference on neural networks* (Vol. 2, pp. 403–409).

Vázquez-Escobar, J., Hernández, J., & Cárdenas-Montes, M. (2021). Estimation of machine learning model uncertainty in particle physics event classifiers. *Computer Physics Communications, 268*, 108100. https://doi.org/10.1016/j.cpc.2021.108100.

Wang, L., Lin, Z. Q., & Wong, A. (2020). Covid-net: A tailored deep convolutional neural network design for detection of covid-19 cases from chest x-ray images. *Scientific Reports, 10*(1), 1–12.

Wang, Z., & Yang, J. (2017). Diabetic retinopathy detection via deep convolutional networks for discriminative localization and visual explanation. *CoRR*. Retrieved from http://arxiv.org/abs/1703.10757

Zech, J. R., Badgeley, M. A., Liu, M., Costa, A. B., Titano, J. J., & Oermann, E. K. (2018). Confounding variables can degrade generalization performance of radiological deep learning models. *CoRR*. ArXiv:abs/1807.00431.

Zhu, L., & Laptev, N. (n.d.). *Engineering uncertainty estimation in neural networks for time series prediction at uber*. https://eng.uber.com/neural-networks-uncertaintyestimatio

A Review of Machine Learning Techniques to Detect and Treat COVID-19 Using EHR Data

Sagnik Nayak, Chirantan Ganguly, and Anil Kumar Gupta

Abstract The COVID-19 pandemic has caused a global emergency, as human life is under constant threat and medical infrastructure is being pushed to the limits. Scientists, engineers, and healthcare experts collaboratively have come up with new innovative technologies to tackle the threat posed by the disease. Artificial intelligence (AI) and machine learning (ML) have played pivotal roles in several technologies being developed. In this study, we aim to review the applications of AI and ML in developing models for diagnosis and clinical outcome prediction of the COVID-19 disease by analyzing data from Electronic Health/Medical Records (EHR/EMR) of patients. In this chapter, we reviewed various ML algorithms used by researchers to extract key features from the EHR necessary to develop predictive models; we also reviewed the performance of various predictive models developed which employed ML to assist decision-making for healthcare professionals. Gaps in the current methodology of acquiring and storing clinical data in a digital format have also been discussed thoroughly. Scientists, clinicians, healthcare experts, policymakers, academics, and ML enthusiasts might find the review useful, as important studies related to the application of ML in the field of medical decision making and alike by studying EHR has been systematically analyzed and critically reviewed, the discussion on the challenges faced by the existing researchers and possible solutions have also been discussed; which should stimulate further research in this domain.

Keywords Machine learning · Artificial intelligence · Electronic health records · Clinical records · EHR · EMR · COVID'19 · SARS CoV-2 · Pandemic · Coronavirus

S. Nayak (✉) · C. Ganguly
Institute of Radio Physics and Electronics, University of Calcutta, Kolkata, India
e-mail: sagniknayak2608@gmail.com

A. K. Gupta
Centre for Development of Advanced Computing (C-DAC), Bengaluru, India

V. Chang et al. (eds.), *Artificial Intelligence and Machine Learning Methods in COVID-19 and Related Health Diseases*, Studies in Computational Intelligence 1023,
https://doi.org/10.1007/978-3-031-04597-4_2

Abbreviations

AI	Artificial Intelligence
ALT	Alanine aminotransferase
ALP	Alkaline phosphatase level
ANN	Artificial Neural Network
AST	Aspartate aminotransferase
AUC	Area Under the Curve
AUROC	Area Under the Receiver Operating Characteristics
BOW	Bag of Words
BMI	Body Mass Index
CBC	Complete Blood Count
CDSS	Computerized clinical decision support systems
CRP	C—reactive protein
CT	Computed Tomography
DL	Deep Learning
DSC	Dice Similarity Coefficient
ECG	Electrocardiography
EHR	Electronic Health Record
GGT	Gamma-Glutamyl Transferase
hs-CRP	High-sensitivity C-reactive protein
ICU	Intensive Care Unit
INR	International Normalized Ratio
IT	Information Technology
LDH	Lactate Dehydrogenase
MAE	Mean Absolute Error
ML	Machine Learning
MRMR	Maximum Relevance Minimal Redundancy
PRISMA	Preferred Reporting Items for Systematic Reviews and Meta-analysis
RF	Random Forest
RT-PCR	Reverse Transcription Polymerase Chain Reaction
SARS	Severe Acute Respiratory Syndrome
SARS-CoV-2	Severe Acute Respiratory Syndrome Coronavirus 2
SVM	Support Vector Machine
XGBoost	Extreme Gradient Boosting

1 Introduction

Humanity has faced numerous virus outbreaks throughout its history, but only a few of them were able to transform the way the world works in a significant manner, the COVID-19 pandemic is certainly one of them. It has been almost a year and a half since COVID-19 was declared a global pandemic (Sohrabi et al., 2021), the World Health Organization (WHO) and its co-operating clinicians and various national authorities around the globe are still fighting against the deadly virus to stop its spread to this date. Not only the health of people has been affected by this pandemic, but there has been a severe impact on the world economy, causing a widespread recession and unemployment (Alankar et al., 2021). As of July 27, 2021, the novel Coronavirus (SARS-CoV-2) disease has spread to over 185 nations, infecting over 194,676,845 people and inflicting 4,167,078 fatalities (WHO Weekly Epidemiological Update; JHU COVID-19 Dashboard).

Clinicians and scientists in the medical field are collaborating to develop technologies that can detect infected patients at various stages, find the best clinical trials, study and predict disease prognosis, research the risks of patients with comorbidities, understand the virus, and develop strategies to limit its spread thereby bringing an end to this ongoing pandemic.

According to recent studies, ML & AI promise new technologies in the field of medicine and healthcare. A variety of healthcare providers use AI and ML since they allow a greater scale-up, faster processing capabilities, and even outperform human abilities in certain healthcare activities (Davenport et al., 2019). As a result, diverse ML and AI technology have been used and are being developed by healthcare enterprises and doctors throughout the world to solve various issues during the pandemic. AI is increasingly used to give decision support for healthcare professionals (Phillips-Wren et al., 2008) who are now working under a lot of stress and unavailability of adequate resources to effectively combat the virus.

The digitalization of the healthcare system and its management in recent years has provided AI and ML technologies with valuable data sources necessary for its proper functioning. We may be able to better understand patients' requirements, the COVID-19 virus, and its transmission by using data from electronic medical records (Dagliati et al., 2021). This insight can be invaluable to assist or fight against the deadly virus.

The aim of this chapter is to offer a complete study and evaluation of the existing corpus of articles researching the application of Machine Learning and Artificial Intelligence technologies to help in the diagnosis and prognosis of COVID-19 infected patients (Fink, 1998).

We explain the search techniques used to identify relevant material to make our analysis transparent and replicable. Following that, we perform a comprehensive assessment of the literature, detailing a range of key aspects of each study. Finally, we go through the problems researchers face in this field and how they can be solved by future research.

2 Methods

A systematic quantitative review was conducted of literature published up until 2021, using the methods outlined by Pickering et al. (2014).

2.1 Search Strategy

The search for relevant literature on the electronic databases was done using the following main keywords:

"COVID-19", "Pandemic", "Electronic Health Records", "Clinical Records", "Machine Learning" and "Deep Learning".

The articles which cite the initial set of included papers were also looked into for finding relevant works. This can be categorized as our secondary search algorithm for finding more relevant literature apart from searching with keywords in electronic databases.

Since ML application in the diagnosis and treatment of COVID-19 crosses multiple disciplines, a literature search was performed in both health and IT databases. IT databases IEEE was searched. We included literature from health-related research databases, such as PLOS Medicine, The Lancet, and PubMed. Databases such as ScienceDirect (Elsevier), Springer, and Nature were used to search for relevant literature. The search did not have a set date range.

2.2 Study Selection

Papers were considered for review if the following inclusion criteria were met:

1. The paper was written in English.
2. The paper discussed the development of ML models to detect COVID-19
3. The paper discussed the development of ML algorithms to predict the clinical outcome of the disease in an individual diagnosed with COVID-19
4. The paper discussed the development of ML algorithms to improve the treatment of COVID-19 patients.
5. The paper was published in a peer-reviewed journal.

The following exclusion criteria were used to filter out papers not relevant to this study:

1. The article was not written in English.
2. The paper did not use ML algorithms to analyze clinical records.
3. The full text of the article was not available (e.g., conference abstracts).
4. The paper was a tutorial or a course material.

This selection criterion is in line with the mental health literature and previous systematic reviews on affective mental health, and it follows the PRISMA guidelines (Liberati et al., 2009). Figure 1 illustrates the flowchart of our search procedure.

The finalized corpus is quantitatively summarized in the findings section. Instead of spanning the breadth of the literature, this method was chosen to highlight the subtleties of the research undertaken. This method also aids in the identification of any gaps in the present literature that future researchers can fill.

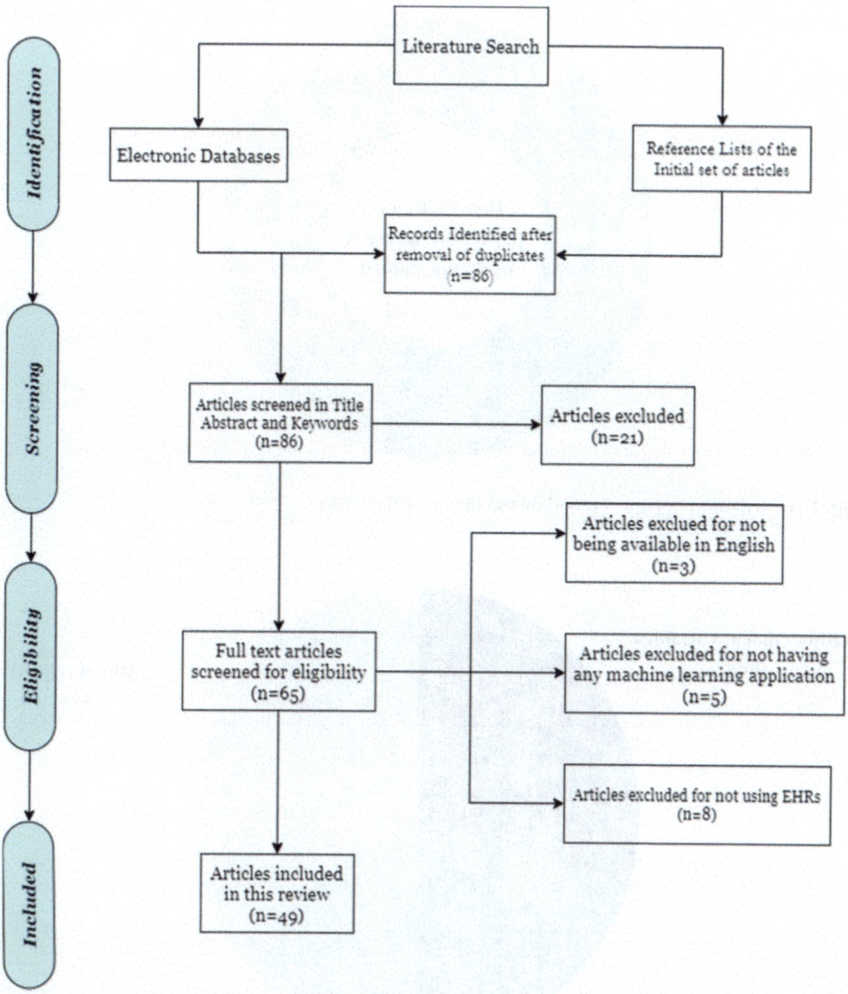

Fig. 1 Procedural flowchart of literature search following PRISMA guidelines

3 Findings

86 studies were identified upon searching the databases which used the following keywords: "COVID-19", "Pandemic", "Electronic Health Records", "EHR", "Machine Learning", "Deep Learning". Following that upon screening and study eligibility review, we ended up with a corpus of a total of 49 papers (Figs. 2, 3, 4 and 5.

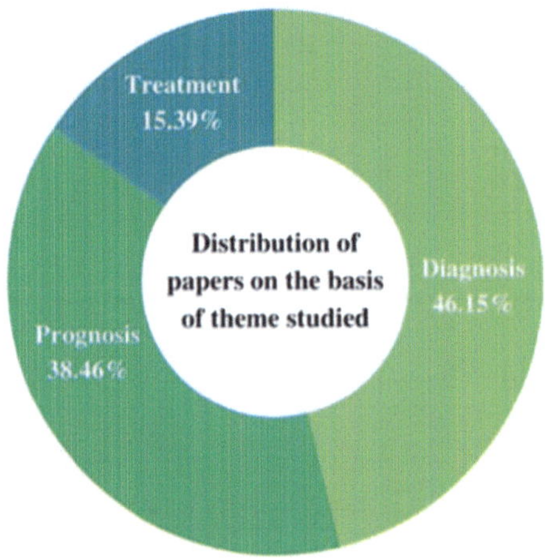

Fig. 2 Distribution of papers based on the theme of the paper

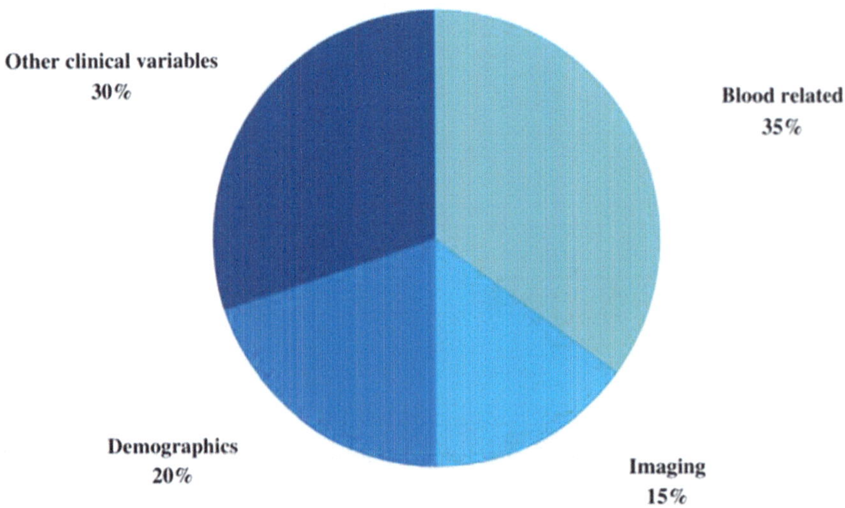

Fig. 3 Distribution of diagnostic models based on the clinical variables used

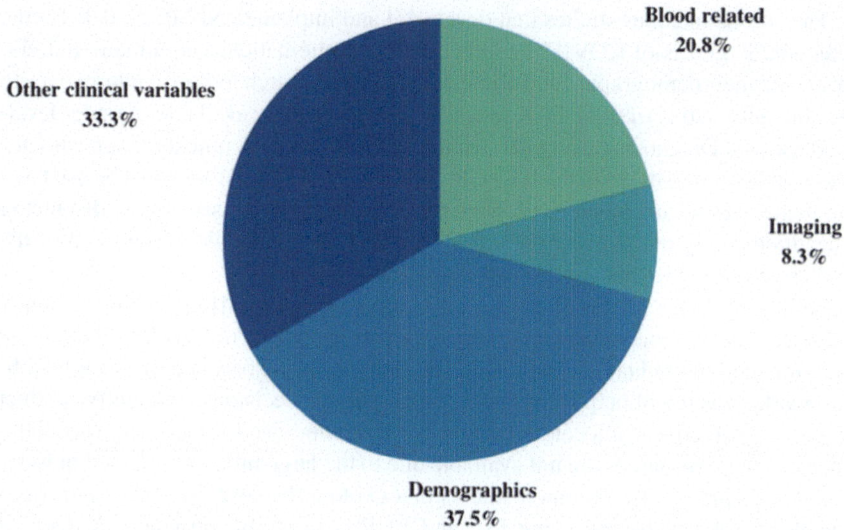

Fig. 4 Distribution of diagnostic models based on the clinical variables used

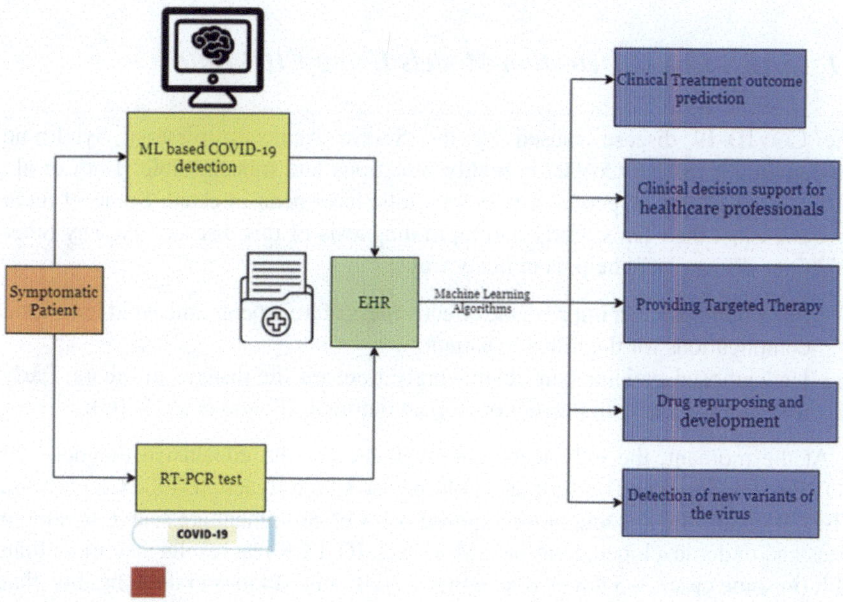

Fig. 5 Using machine learning (ML) with electronic health record (EHR) data in managing the COVID-19 pandemic

The corpus includes studies that developed and implemented ML models for the detection/diagnosis of COVID-19 infection in symptomatic/asymptomatic individuals using their demographic and clinical data. This research looks for alternatives to the clinically authorized RT-PCR test, which is not widely available in underdeveloped nations. Diagnosing using ML techniques also provides quicker diagnosis (in a matter of few seconds) than the RT-PCR test (3–5 days). Therefore it can be used as a supplementary testing method, allowing patients to adopt necessary social-distancing precautions early on, slowing the virus' transmission. A detailed discussion of the studies pertaining to this application is given in Sect. 3.1.

Following that, a review of articles advocating the use of ML algorithms to study the disease development and predict the treatment outcomes of COVID-19 infection in an infected individual is conducted. This study of prognosis is critical for healthcare workers as it will help them make critical clinical decisions, efficiently conduct triage, and effectively allocate resources to those who need it the most, especially since adequate resources are not available due to the large influx of infected individuals worldwide. A discussion on studies that explore the development of improved treatments for the disease using ML and EHR data is also provided. A detailed overview of the papers reviewed is presented in Sect. 3.2 of this study.

3.1 SARS-CoV-2 Detection Models Using EHR Data

The COVID-19 disease caused by the Severe Acute Respiratory Syndrome Coronavirus-2 (SARS-CoV-2) is highly infectious and transmissible (Lotfi et al., 2020), and millions of people across the globe have been infected. Some of them have even lost their lives. Early and rapid diagnosis of this disease, like any other infectious disease, will help us in many ways:

i. Early diagnosis can improve the effectiveness of treatments and avoid long-term complications for the infected patient.
ii. Undiagnosed patients can unknowingly transmit the disease to others. Early diagnosis can help to prevent or stop an outbreak (Banoo et al., 2007).

At the moment, the only technique available for the conclusive diagnosis of COVID-19 is Reverse Transcriptase Polymerase Chain Reaction (RT-PCR) testing. RT-PCR diagnostic kits and other required types of equipment are scarce in remote areas and underdeveloped countries. Moreover, RT-PCR test results take more than 24 h (in some cases 3–5 days); due to this scarcity in test kits and the long time that is required to get the test results, a lot of crucial time is lost, which could be used in providing treatment at early stages. Scientists have thus been trying to develop effective screening and testing techniques that can faithfully detect COVID-19 patients from easily accessible data and can give results in lesser amounts of time.

Machine learning (ML) is a set of computer algorithms that improve themselves over time as a result of their utilization of data and experience. ML algorithms construct statistical models based on available data, known as "training data," in order to make predictions without being explicitly programmed to do so (El Naqa et al., 2015). Due to these reasons, machine learning technologies are being seriously considered by the researchers to use in clinical decision-making. CDSS, or computerized clinical decision support systems, represents a paradigm change in today's healthcare. CDSS is utilized to help physicians with their difficult decision-making. Since its introduction in the 1980s, CDSS has evolved rapidly. EHRs and other automated healthcare processes are now used as data sources for CDSS models. This development has been possible due to the growing global usage of sophisticated electronic medical records and the development of various ML algorithms (Sutton et al., 2020).

ML models might be used to extract features from existing data to determine which characteristics have the highest association with being infected with COVID-19, allowing us to investigate new and faster approaches to identify COVID-19 in patients.

In recent studies, ML models are being explored and developed to detect COVID-19 infection in individuals from readily available data in their Electronic Medical Records. Most researchers have used demographic data, symptoms of individuals, and some easily accessible laboratory test results (such as routine blood tests) to build their ML model.

In their study, Joshi et al. (2020) developed an ML algorithm that uses just the complete blood count (CBC) and gender to predict the COVID-19 PCR test positivity. The authors trained a logistic regression model based on historical data from EHRs of patients in the emergency department of a single medical institute and tested it using data from other institutes across the world. A sensitivity of 93% and c-static 78% were observed for the developed COVID-positivity prediction algorithm.

In a similar study done by Banerjee et al. (2020), an ML model was developed using ANN to identify SARS-CoV-2 positive patients from full blood counts data without knowledge of symptoms or history of the individuals. The dataset used in this study was anonymized data of patients being treated at a hospital in São Paulo, Brazil, and also those who had performed RT-PCR and other tests at that hospital during a visit. The ANN model developed by them was able to classify COVID-positive patients with an accuracy of 89% for the community ward and 90% for the regular ward, respectively.

In another study done by Soltan et al. (2020), routinely obtained healthcare data was used to develop and assess early-detection models for COVID-19 screening among patients entering the emergency department and the subgroup being admitted to a set of four hospitals. The health data extracted included regular blood test results, blood gas testing, vital signs, and RT-PCR test reports. The screening model developed used linear (logistic regression) and non-linear (random forest and XGBoost) classifiers. The test set was generated from an 80:20 stratified train-test split of the dataset and balanced equally with controls (approximately 50%). The early detection model achieved an AUROC of 0.939 with a sensitivity of 90%.

Kukar et al. (2021) worked on developing another COVID-19 screening ML model, which uses regular blood test data of 5333 patients with different viral diseases and 160 COVID-19-positive individuals collected from the Department of

Infectious Diseases, University Medical Centre Ljubljana (UMCL), Slovenia. The cross-validated AUC of the generated model was 0.97. According to the XGBoost algorithm's feature significance rating, the five most relevant routine blood parameters for COVID-19 diagnosis were: Mean Corpuscular Hemoglobin Concentration, eosinophil count, albumin, INR (International Normalized Ratio), and prothrombin activity percentage.

A diagnostic model for COVID-19 in the inpatient environment employing ML algorithms was developed and tested by Goodman-Meza et al. (2020). The algorithm was created to act as a screening tool in hospitals when testing was limited or unavailable. Data from EHRs at the UCLA Health System in California was used retrospectively. The research covered all individuals who had opted for RT-PCR testing and had a set of other laboratory results. The AUROC of the model was 0.91, and the model's sensitivity was 0.93, and its specificity was 0.64. The diagnostic measures of the machine learning system were equivalent to RT-PCR testing.

Plante et al. (2020) developed and evaluated another ML model to rule out COVID 19 in adults in emergency rooms using just standard blood test results for building their model. Clinical data from emergency departments of 66 hospitals in the US was utilized to carry out this study before and during the pandemic. The model was tested using 2183 PCR-confirmed cases from 43 hospitals during the pandemic and 10,000 pre-pandemic individuals as negative controls from the same hospitals. The model was developed using the gradient boosting technique provided in XGBoost. The AUROC for training and external validation of the model was 0.91 with a sensitivity of 95.9% and specificity of 41.7%.

Langer et al. (2020) developed an ML algorithm that uses only primary health data to predict COVID-19 positivity in patients with influenza-like symptoms. The clinical data used to train and evaluate the algorithm was the data obtained upon admission of patients to the emergency department of one of Milan's (Italy) most important hospitals. The dataset included parameters such as age, gender, type of comorbidities present, reported symptoms, and medication currently being taken. Artificial Neural Networks were used in developing the model. When evaluated, the model resulted in an accuracy of 91.4% with 94.1% sensitivity and 88.7% specificity. This research shows that fundamental clinical data may be adequate for appropriately trained ML algorithms to accurately predict SARS-COV-2 positivity and negativity in patients with influenza-like symptoms.

D'Ambrosia et al. (2020) in their study, attempted to build and clinically validate an adaptive and personalized diagnostic model to aid physicians in determining COVID-19 positivity based on their symptoms and some imaging and test data (Chest Radiography, Computed tomography, RT-PCR, Antibody). The data from the Center for Systems Science and Engineering at Johns Hopkins University was used to conduct this study. Various ML algorithms were used for performing the diagnostic classification task; of all the algorithms studied, Logistic regression resulted in the highest accuracy of 90.6% with 100% specificity.

Khanday et al. (2020) in their study used textual clinical reports to diagnose COVID-19. Data used in this study is available in a repository available on GitHub, which has data of 212 patients stored; these patients have shown influenza-like symptoms. The dataset consisted of about 24 attributes about the patients' demographics, symptom progression, and other clinical data such as O_2 saturation percentage, leukocyte count, neutrophil count, lymphocyte count, and other relevant clinical notes. ML techniques were used to divide this dataset into four categories. Term frequency/inverse document frequency, Bag of words (BOW), and report length methods were used for selecting important features. An ensemble of machine learning classifiers was given the features selected. Logistic regression and Multinomial Naive Bayes outscored other ML algorithms with 96.2% testing accuracy.

Abdel-Basset et al. (2021) presented in their study a semi-supervised approach for classifying COVID-19 lung infections from only a few labeled lung scans. A few-shot segmentation (FSS) deep-learning model is developed to study and accurately segment CT-scan results. The CT scan datasets which were used by the researchers for model evaluation were published by the Italian Society of Medical Interventions and Radiology. The model developed by the researchers outperforms other contemporary studies with COVID-19 segmentation Dice Similarity Coefficient (DSC) of 0.789, with 80.3% sensitivity and 98.6% specificity. The CT-scan data is very often available and is maintained in EHR systems of hospitals and is thus readily available. This image segmentation model can be very well used to classify COVID-19 patients from others using CT-Scans instead of RT-PCR tests.

Dhiman et al. (2021), a high accuracy developed deep learning model to identify infected patients using chest X-rays. Eleven different CNN models were developed to perform the task and were evaluated. The data used for developing and evaluating this model were taken from GitHub open-source repositories and a Kaggle dataset. The accuracy of ResNet101 CNN plus the J48 decision tree approach yielded the best accuracy among all the models developed. X-ray images of the chest are easily attainable even in the most remote health infrastructure, so models such as this one can act as a great boon for places where resources are limited. Further investigation and validation with a more robust dataset will further verify the effectiveness of the model.

Brinati et al. (2020) used hematochemical variables from regular blood tests to construct two ML categorization algorithms. Parameters such as white blood cell counts, platelets, CRP, AST, ALT, GGT, ALP, LDH plasma levels were used to train and test the model. The study's data came from 279 individuals who were hospitalized to a hospital in Milan, Italy, with COVID-19 symptoms and were tested with the RT-PCR test. 177 of these people tested positive for COVID-19, whereas 102 tested negative. The two machine learning models were created using the Random Forest classifiers, and their accuracy ranged from 82 to 86% and their sensitivity between 92 and 95%. In addition, a *Web-based tool* for clinical use and study assessment was made accessible.

All the studies mentioned above developed ML models widely implemented in areas where RT-PCR tests and specialist facilities are scarce. Various data types are used by the studies discussed above; models which suit the EHR maintained in various healthcare facilities can be used by them effectively. This technique can also be used as a secondary testing mechanism by hospitals and testing centers across the globe to get faster results (in a matter of minutes) than the standard RT-PCR test, which typically takes a few days.

A summary of all the articles studied has also been presented in a tabular format in Table 1 for easier access.

Table 1 ML and AI technology in SARS-CoV-2 screening, using digital clinical records

Reference	ML/AI method	Type of data	No. of patients	Results
Joshi et al. (2020)	Logistic regression	Complete Blood Count (CBC) and Gender collected from historical EHR	Train: 390 (357 PCR negative, 33 PCR positive) Test: 2,387 (2052 PCR negative, 335 PCR positive)	Sensitivity 93% C-static: 78%
Banerjee et al. (2020)	Artificial neural network	Full blood counts collected from EHRs	598 (517 PCR negative, 81 PCR positive)	An accuracy of 89–90% was achieved
Soltan et al. (2020)	Logistic Regression, Random Forest, XGBoost	Blood tests, blood gas testing, vital signs, and results of RT-PCR	114,957 patients before December 2019 formed the COVID negative cohort 437 patients who tested RT-PCR positive formed COVID positive cohort	AUROC: 0.939 Sensitivity: 90%
Kukar et al. (2021)	XGBoost	Regular blood test data	5,493 (5,333 COVID negative, 160 COVID positive)	AUC: 0.97 Most relevant blood parameters for COVID-19 diagnosis: MCHC, eosinophil count, albumin, INR, and prothrombin activity percentage

(continued)

Table 1 (continued)

Reference	ML/AI method	Type of data	No. of patients	Results
Goodman-Meza et al. (2020)	Random forest, logistic regression, support vector machine, multilayer perceptron (neural network), stochastic gradient descent, XGBoost, and AdaBoost	EHR data Age, gender, hemoglobin, red blood cell count, absolute neutrophil, absolute lymphocyte, absolute eosinophil and absolute basophil counts, the neutrophil to lymphocyte ratio, C-reactive protein, ferritin, and LDH	1455 cases remained (1273 PCR negative and 182 PCR positive)	AUROC: 0.91 Sensitivity: 0.93 Specificity: 0.64 Features that provided most of the information to the model were: C-reactive protein and LDH
Plante et al. (2020)	XGBoost	Standard blood tests	Training: 2183 PCR-positive; 10,000 prepandemic as negative controls Testing: 1020 PCR-confirmed cases and 171,734 pre-pandemic negative controls	AUROC: 0.91 Sensitivity: 95.9% Specificity: 41.7%
Langer et al. (2020)	ANN	Gender, type of comorbidities present, reported symptoms, and medication currently being taken	199 (75 PCR negative, 124 PCR positive)	Accuracy: 91.4% Sensitivity: 94.1% Specificity: 88.7%
D'Ambrosia et al. (2020)	Logistic regression	Chest Radiography, Computed tomography, RT-PCR, Antibody	55 (38 RT-PCR positives, 17 RT-PCR negatives)	Accuracy: 90.6% Specificity: 100%
Khanday et al. (2020)	Logistic Regression; Multinomial Naive Bayes	Demographics, symptom progression and clinical data (O_2 saturation percentage, leukocyte count, neutrophil count and lymphocyte count)	212	Accuracy: 96.2%

(continued)

Table 1 (continued)

Reference	ML/AI method	Type of data	No. of patients	Results
Abdel-Basset et al. (2021)	Few-shot segmentation (FSS) Deep-Learning model	Lungs CT-scan data	60	DSC:0.789 Sensitivity: 0.803 Specificity: 0.986 Sα: 0.834 Eφ: 0.908 MAE: 0.065
Dhiman et al. (2021)	Convoluted Neural Network (CNN) and Decision Tree	Chest X-ray	100 (50 COVID-19 positive, 50 COVID-19 negative)	Accuracy: 100%, Recall: 100%, Specificity: 98.89%, Precision: 100%, and F1-score: 100%
Brinatiet al. (2020)	Random Forest classifier	Routine blood tests (parameters: white blood cell counts, platelets, CRP, AST, ALT, GGT, ALP, LDH plasma levels)	279(177 RT-PCR positives, 102 RT-PCR negatives)	Accuracy: 82–86% Sensitivity: 92–95%

3.2 *SARS-CoV-2 Prognostic Models Using EHR Data*

Prediction algorithms that assess the likelihood of people experiencing serious clinical outcomes as a result of COVID-19 infection might help doctors prioritize patients when allocating limited healthcare resources available. Prognostic models ranging from rule-based grading systems to sophisticated machine learning models have been suggested, developed, and published by researchers across the globe in an attempt to guide the public health response to the COVID-19 crisis and ultimately save lives.

Das et al. (2020) developed such a prognostic model utilizing ML algorithms to predict mortality among confirmed COVID-19 patients in South Korea, employing publically accessible data from Korea Centers for Disease Control and Prevention (KCDC) that included demographic and exposure parameters such as age, sex, province, and exposure. The model was also used to develop an open-source online decision-making tool. Five machine learning methods were used to predict mortality for confirmed CoVID-19 patients (n = 3,524). (Logistic regression, support vector machine, K-Nearest Neighbor, random forest, and gradient boosting). The logistic regression technique outperformed the others in terms of classification (AUROC = 0.830). Prospective stakeholders, such as health practitioners and legislators, might use this method, in addition to other strategies, to triage patients at the community level.

Schöning et al. (2021) advocated using ML models on regular laboratory results to aid in the automated detection of SARS-CoV-2 patients at risk of serious clinical consequences (admission to ICU or death). This can help in patient triage, healthcare resource allocation, and clinical outcome improvement. They created an ML model using Support Vector Machine, trained using demographic data, medical history, and laboratory results from 198 patients admitted to a group of hospitals in Bern, Switzerland, who tested positive for SARS-CoV-2. The validation cohort consisted of 459 patients from the same institution. The model had an AUROC of 0.85, and the strongest predictive factors for clinical outcomes were sex, C-reactive protein, sodium, hemoglobin, glomerular filtration rate, glucose, and leucocytes around the time of first positive testing.

Yadaw et al. (2020) sought to develop an accurate COVID-19 mortality prediction algorithm and find the clinical characteristics that were most predictive of this result using machine learning techniques. ML algorithms were used to predict mortality to study clinical data from 5051 COVID-19 patients (416 of whom expired) from the Mount Sinai Health System in New York City, USA. When applied to test datasets of retrospective ($n = 961$) and prospective ($n = 249$) patients, the COVID-19 mortality prediction model created using the XGBoost algorithm demonstrated good accuracy ($AUC = 0.91$). The most significant clinical characteristics for mortality prediction were the patient's age, minimum oxygen saturation during the length of their clinical interaction, and kind of patient encounter (inpatient vs. outpatient and telehealth visits).

In a study by Rechtman et al. (2020), they attempted to find variables accessible immediately after the initial clinical examination to predict COVID-19 mortality. A retrospective analysis was done on 8770 laboratory-confirmed SARS-CoV-2 patients from a group of 53 hospitals in New York City. ML algorithms were used to analyze demographic, clinical data, and comorbid characteristics of the patients from their EHRs. Older age, male sex, greater BMI, higher heart rate, higher respiratory rate, lower oxygen levels, and chronic kidney disease were found to be the most correlated with COVID-19 mortality using logistic regression. When gradient-boosting machine learning was used to predict COVID-19-related mortality using the previously stated factors, it resulted in an AUC of 0.86. The results of this study show that objective measures available at the time of diagnosis are good predictors of COVID-19 outcome. The findings from this study might help us to improve healthcare delivery through fast response techniques.

Fernandes et al. (2021) designed multifunctional algorithms to predict the likelihood of COVID-19 patients acquiring critical conditions in their research. For this investigation, a total of 1040 patients with a positive RT-PCR diagnosis for COVID-19 were followed in a hospital in Sao Paulo, Brazil, of which 288 (28%) had a severe prognosis, i.e., ICU admissions, use of mechanical ventilation, or death. Five machine learning algorithms (artificial neural networks, extra trees, random forests, catboost, and extreme gradient boosting) were trained and tested using routinely using the gathered laboratory, clinical, and demographic data from the previously mentioned individuals. All of the algorithms performed quite well in terms of prediction (average AUROC of 0.92, sensitivity of 0.92, and specificity of 0.82). The ratio of lymphocytes per C-reactive protein, C-reactive protein, and Braden Scale were the three most essential parameters for the multipurpose algorithms.

Gao et al. (2020) designed a COVID-19 mortality risk prediction model that employs patients' clinical data on admission to stratify patients by mortality risk. The model can forecast physiological decline and death up to 20 days in advance. Logistic Regression, Support Vector Machine, Gradient Boosted Decision Tree, and Neural Network are the four ML algorithms used to create the ensemble model. For model training and validations, epidemiological, demographic, clinical, laboratory, radiological, and outcome data from EHRs utilizing identical data collecting forms on the first day of admission of 2160 patients diagnosed with COVID-19 in Wuhan, China, were used. An internal validation cohort and two external validation cohorts were used to validate the model, with AUCs of 0.9621, 0.9760, and 0.9246, respectively.

In a study done by Karthikeyan et al. (2021), they present an ML model for predicting COVID-19 mortality using only regular blood test data. A strong combination of five blood characteristics, namely: lymphocytes, neutrophils, high-sensitivity C-reactive protein (hs-CRP), lactate dehydrogenase (LDH), and age, has been found to predict death with almost 96% accuracy. Several ML models were trained and tested, namely neural networks, logistic regression, XGBoost, random forests, SVM, and decision trees, to determine the model which consistently yields high prediction accuracy across the disease's stages. The best algorithm was the one that employed XGBoost for feature selection and neural network classification, and it was able to forecast mortality with 90% accuracy almost 16 days before the occurrence.

In their study, Ryan et al. (2020) developed and evaluated a machine learning-based model that accurately predicts mortality in COVID-19, pneumonia, and mechanically ventilated patients. Data from 53,001 ICU patients, including 9166 pneumonia patients and 25,895 mechanically ventilated patients, were retrospectively analyzed and used to train the ML model (employing XGBoost). A dataset of 114 individuals who tested positive for SARS-COV-2 was also obtained from a community hospital and used for analysis. For mortality prediction of mechanically ventilated patients, the XGBoost predictor resulted in an average AUROC of 0.7875; for pneumonia patients, an AUROCs of 0.7885 was observed. When evaluated on the COVID-19 dataset, it achieved average AUROCs of 0.885 for death prediction of patients was obtained.

Parchure et al. (2020) in their study proposed and demonstrated the use of a machine learning (ML) model for predicting near-term in-hospital death among COVID-19 patients. Data from nursing flowsheets (e.g., vital signs, breathing pattern); demographics (e.g., age, sex); appropriate hospital organizational factors (e.g., admission type, source of admission); relevant laboratory findings, and ECG derived characteristics from a cohort of 567 COVID-19 patients was obtained from the Mount Sinai Health System and was used to perform the study. The random forest model was trained on 70% of the randomly selected cohort (training set) and was tested on the remaining 30% (the test set). The RF classifier resulted in an AUROC of 85.5%, a sensitivity of 87.8%, a specificity of 60.6%, and an accuracy of 65.5%.

Shiri et al. (2021) employed demographics data and medical history, laboratory test results, visual rating by radiologists, and lung/lesion radiometric characteristics derived from chest CT images to develop a predictive model for survival prediction (alive or dead status) of COVID-19 patients. A total of 152 patients were studied (data of 106 patients was used for training of the model, and the remaining data from 46 patients was used for testing). The maximum relevance minimal redundancy (MRMR) and XGBoost algorithms were utilized for feature selection and classification. Cancer comorbidity, consciousness, oxygen saturation, and blood urea nitrogen level were all found to be strongly linked to a person's mortality result. The overall prognostic model resulted in an AUC = 0.95, accuracy = 0.88, sensitivity = 0.88 and specificity = 0.89.

All the studies discussed here illustrate how ML models for predicting clinical outcomes may be developed and deployed using a limited set of easily accessible characteristics, which will aid in efficient triage and, as a result, potentially decrease mortality rates in patients.

A summary of all the articles studied has also been presented in a tabular format in Table 2 for easier access.

4 Treating COVID-19 Using Machine Learning and EHR Data

4.1 Repurposing of Commercially Available Drugs and Vaccines for the Treatment of COVID-19

At the beginning of the global outbreak of COVID-19 disease across the globe, there was a looming state of crisis as there were no tested and approved drugs for this disease. Drug design and testing protocols are very complex; it is a very elaborate and time-consuming process. Therefore, many researchers used ML algorithms to predict the efficacy of the already designed drugs and commercially available in treating the COVID-19 disease, which was initially created for treating other diseases.

Mohapatra et al. (2020), in their study, predicted the therapeutic efficacy of ten FDA approved drugs in treating the COVID-19 disease using the Naïve Bayes algorithm. The developed model was able to predict drug efficacy with an accuracy of 73%. They were able to find three drugs among the ten studied to be potentially effective in treating COVID-19 disease, of which Amprenavirwas found to be the most effective drug-based on specialized criteria.

In a similar work done by Hsieh et al. (2020), they studied 3635 drugs used in COVID-19 treatment. A deep neural network model was created to find out the drugs that had the best response in treating the disease. Various parameters such as the patient's clinical trial history and genetic profiles were extracted from the patients'

Table 2 ML and AI applications in predicting and forecasting clinical outcomes of SARS-CoV-2 using EHRs

Reference	ML/AI Method	Type of data	No of Patients	Results
Das et al. (2020)	Logistic Regression	Demographics and exposure data (age, sex, province, and exposure)	3,524 (74 deaths)	AUROC: 0.830 Age was the most important predictor, followed by exposure, sex, and province
Schöning et al. (2021)	Support Vector Machine	Demographic data, medical history, and laboratory results collected at the time of first positive testing	Train: 198 patients Test: 459 patients	AUROC: 0.85 Strongest predictive factors for clinical outcomes: gender, C-reactive protein, sodium, haemoglobin, glomerular filtration rate, glucose level, and leucocytes count
Yadaw et al. (2020)	XGBoost	Clinical Data	Train: 5051 COVID-19 patients (416 of whom expired) Test: Retrospective data (961 patients) and prospective data (249 patients).	AUC: 0.91 Most predictive factors: age, minimum oxygen saturation, and kind of patient encounter (inpatient vs outpatient and telehealth visits)
Rechtman et al. (2020)	Logistic Regression, gradient boost algorithm	Demographics data, clinical data, and comorbid characteristics collected from EHRs	8770. AUC of 0.86.	AUC of 0.86 Factors that increased mortality were Older age, male sex, greater BMI, higher heart rate, higher respiratory rate, lower oxygen levels, and chronic kidney disease

(continued)

Table 2 (continued)

Reference	ML/AI Method	Type of data	No of Patients	Results
Fernandes et al. (2021)	Artificial neural networks, random forests, catboost, and XGBoost	Routinely gathered laboratory test results, clinical and demographic data.	1040 patients, 288 (28%) had a severe prognosis (ICU admissions, use of mechanical ventilation, or death).	AUROC = 0.92, Sensitivity = 0.92, Specificity = 0.82 The ratio of lymphocytes per C-reactive protein, C-reactive protein, and Braden Scale were the three most essential parameters for the multipurpose algorithms
Gao et al. (2020)	Logistic Regression, SVM, Gradient Boosted Decision Tree, and Neural Network.	Demographic, epidemiological, clinical, laboratory, radiological, and outcome data from EHRs	2160 patients	An internal validation cohort and two external validation cohorts were used to validate the model, with AUCs of 0.9621, 0.9760, and 0.9246, respectively
Karthikeyan et al. (2021)	Neural networks, random forests, logistic regression, XGBoost, SVM, and decision trees	Blood test data from EHRs	2,779 patients	XGBoost feature importance and neural network classification performed the best; they had the ability to predict mortality with 90% accuracy as early as 16 days before the event The following 5 parameters predicting death with 96% accuracy: neutrophils, lymphocytes, lactate dehydrogenase (LDH), high-sensitivity C-reactive protein (hs-CRP), and age
Ryan et al. (2020)	XGBoost	Raw EHR data	53,001 ICU patients; 9166 pneumonia patients; 25,895 mechanically ventilated patients; 114 COVID-19 patients	The following results were obtained : Time Window 12 hr 24 hr 48 hr 72 hr Ventilator AUROC 0.82 0.81 0.77 0.75 Pneumonia AUROC 0.87 0.78 0.77 0.734 COVID-19 AUROC 0.91 0.90 0.86 0.87

(continued)

Table 2 (continued)

Reference	ML/AI Method	Type of data	No of Patients	Results
Parchure et al. (2020)	Random Forest classifier.	Demographics; relevant hospital administrative variables (eg, admission type, source of admission); data from nursing flowsheets (eg, vital signs, respiratory pattern); relevant laboratory results and ECG derived variables	567 patients (train—70%, test—39%)	AUROC = 85.5 %, Sensitivity = 87.8%, Specificity = 60.6%
Shiri et al. (2020)	Maximum Relevance Minimal Redundancy (MRMR); XGBoostThe overall prognostic model resulted in an AUC = 0.95, accuracy = 0.88, sensitivity = 0.88 and specificity = 0.89	Demographics and history, laboratory tests, visual rating by radiologists and lung/lesion radiomic characteristics derived from chest CT images	152 patients (106 for training/validation and 46 for testing)	AUC = 0.95 Accuracy = 0.88 Sensitivity = 0.88 Specificity = 0.89 Cancer comorbidity, consciousness, oxygen saturation, and blood urea nitrogen level were all found to be strongly linked to a person's mortality

electronic health records to study their disease prognosis and how the drug affected the course of the disease. The overall ranking model gained an AUROC of 89%, and the best 22 drugs were identified. Of these 22 drugs identified, Azithromycin, Hydroxychloroquine and Atorvastatin yielded the best results.

Gupta et al. (2021) developed a multivariable logistic regression model trained using EHR data obtained from patients admitted with COVID-19 in hospitals in the United States. The developed model suggests that the patients with COVID-19 who have been receiving statins have a much lesser in-hospital mortality rate than those who are not.

Analysis of the data obtained from Mayo Clinic EHR by Pawlowski et al. (2021) interestingly reveals that individuals who had received various commercially available vaccines (such as the vaccines against influenza, hepatitis A, hepatitis B, pneumococcal conjugate, measles-mumps-rubella, Haemophilus influenza or varicella) within 5 years before the pandemic had decreased rates of test-positivity for COVID-19.

Therefore, these EHR-based computational studies bring forward valuable insights into how the virus reacts against various drugs. This will help us identify which existing drugs have the potential to treat the COVID-19 disease, but it will also help in the development of new drugs to treat the COVID-19 disease.

4.2 *Development of Targeted Medical Therapy Techniques*

With the surfacing of the numerous various variants of the COVID-19 disease, a varied response to the developed medication and vaccines is expected from the different variants. This varied responses might be manifested as either an improved therapeutic outcome by the developed drugs; however, it may also lead to adverse reactions produced by the drugs. Thus it is essential to track and predict this varied response to the drugs by the different variants at early stages—which will help us target specific medical therapies to specific COVID-19 variants, thereby minimizing risk. For attaining this objective genomic sequencing of the COVID-19 in hospitalized patients must be performed and added to their EHR, thus making it possible to track response to various drugs by patients infected with various variants.

Moreover, with ML and EHR data deployment, clinical decision support systems may be created, which would help notify physicians by predicting from preset conditions if patients would need specific medical therapies such as supplemental oxygen or convalescent plasma therapy. The development of such systems would potentially help in achieving better therapeutic outcomes.

5 EHR and Machine Learning in Studying Mutations in the Virus

The virus which causes coronavirus disease (SARS_CoV-2) is a positive-sense, single RNA virus. Like other RNA viruses, SARS-CoV-2 has high mutation rates. As the virus evolves, the transmissibility and infectivity may get altered, leading to the developed drugs being rendered less effective against some future mutation of the virus. This can also lead to rendering some treatments methods less effective than others; some changes in the prognosis and symptoms of the disease might also be noticed.

Analyzing the prognostic and symptom patterns from the clinical data of patients and prospective patients, looking for a significant pattern of changes from the symptoms observed in a particular locality can thereby indicate the possible creation of mutants (Mahase, 2021). This would be crucial, as by closely studying the clinical data, we can track the mutation near about its very inception; after that, we can take necessary measures to stop the spreading of the mutated virus. This can save us from the uncertain spread of potential virus mutants resistant to the currently developed drugs. For carrying this out, exploratory data analysis on the clinically obtained data can be a great option.

EHR systems should also track the genetic variants detected in patients infected with SARS-CoV-2; this will help us identify clusters of specific concerning variants in certain localities. This would prove crucial as more resources can be directed towards those localities, potentially resulting in better therapeutic outcomes, and travel restrictions may be implemented in such areas to limit further spread of the concerning variants.

6 Early Detection of Future Pandemics

The importance of early detection to an outbreak of a new disease is immense. The COVID-19 pandemic very well demonstrated this as the late detection of the disease has scaled the disease up to a worldwide pandemic. Quick response and critical mitigation strategies are crucial to prevent the spread of the virus in its initial stages to stop its global spread, sky-rocketing of cases, and thereby prevent the ruckus in which we are in right now.

The COVID-19 pandemic severely affected the healthcare infrastructure of almost every country in the world, and from this experience, the global community should work collaboratively for early detection of novel diseases by using data from various sources, such as electronic health records, Internet browsing history of symptoms of disease in a particular locality, and other sources. Data Analysis and Machine

Learning algorithms might be used to classify such data and giving us early indications of the potential emergence of novel diseases, following which widespread genomic sequencing of people with like symptoms may be done to detect any new virus.

7 Challenges and Future Direction

The COVID pandemic has made the pivotal role that ML and AI will play in the healthcare industry in the future pretty clear. However, to explore this field further, a unified framework for sharing and exchanging digital clinical records is required, and proper data protection regulations must be devised to ensure the privacy of the individuals (Dagliati et al., 2021).

Although the advent of machine learning in clinical decision-making facilitates efficient handling of the pandemic for health-workers and stakeholders, a better standard of digital clinical records is required for further development of this domain. EHR utilization has increased dramatically—from 2008 to 2015, the usage of any form of EHR increased from 9% to 96% for hospitals and from 17 to 78% for physician offices (Report To Congress on Health IT Progress, 2016); however, only a tiny percentage of them maintain their EHR systematically and comprehensively. Since there has been a rise in the usage of EHR for helping in decision support, this large, heterogeneous, incomplete, and noisy form of data storage has to be done away with, and new techniques have to be explored for systematic but easy filling of the health records by the clinicians so that this data can be effectively used for developing computerized decision support systems (CDSS) (Kim et al., 2019).

A lot of research was also plagued by small sample sizes and localized samples due to the unavailability of data in EHRs; this is a major challenge as small sample size and the high number of input variables bear an intrinsic risk of over-fitting in machine learning models (Wolff et al., 2019).

The most difficult challenge for applying AI in various healthcare fields is not whether the technologies are sophisticated enough to be useful, but it is securing its acceptance in everyday clinical practice. In order to be extensively utilized, AI systems must be acknowledged by authorities and sufficiently integrated with EHR systems. These obstacles will be overcome in the end, but they will take considerably longer than the maturation of the technologies themselves (Davenport et al., 2019).

8 Conclusion

In this systematic review, we assembled the current literature that utilized ML methods in studying and extracting valuable information from Electronic Health Records and thereby help in detecting, performing prognosis, and treating the viral infection caused by SARS-CoV-2. Following this, we also discussed possibilities of using clinical data and ML methods to detect the onset of future novel diseases to take control of it effectively, not letting cases sky-rocket across the globe.

Our review of the literature clearly suggests that ML-based tools for diagnosis of COVID-19 are highly accurate in detecting the presence of the virus using digitalized clinical data and regular laboratory tests, which are available for and can be maintained for cheap, while also providing fast results, and therefore it has tremendous potential and should be definitely employed as a secondary COVID-19 testing procedure where access to labs and equipment are not readily available.

The prognostic tool works well in performing risk-based stratification of infected patients, which can help healthcare professionals in allocating resources in an intelligent manner. This can help us save the lives of individuals in dire need of special care by informing the doctors well before situations worsen. Clinical data can also be used to cater specific treatment to different individuals suffering from various variants of COVID-19, which might have further disease development and symptoms. This can be done by classifying symptoms according to various variants. Thus robust clinical decision support systems can be developed by using Machine Learning algorithms on clinical data, which can assist us in efficiently managing the pandemic with the efficient allocation of scarce resources.

Similarly, the observation of symptoms that cannot be categorized to other discovered COVID-19 variants suggests infections caused by a new mutation of SARS-CoV-2, which we can further study by genomic sequencing. This mechanism can also help us in the future to study the surfacing of new viruses and infectious diseases by comparing the symptoms to other already discovered diseases, thereby helping us in its rapid detection and thereby control its spread across the globe.

Thus, in conclusion, we can confidently say that the potential for applying Machine Learning techniques in the study of infectious diseases is widespread. EHRs are the naturally suited data source to train and test such algorithms. EHR at its current state needs a lot of improvements as they are not maintained in a manner conducive to the building of ML models and AI technologies. Data sharing between EHRs of different hospitals and public health entities can make clinical data more robust and therefore improve disease tracking and contact tracing to limit spread.ML and AI-based approaches can improve treatment outcomes for patients and advance the scientific understanding needed to combat the COVID-19pandemic and any other pandemics of the future.

9 Limitations

The corpus covered in this study is far from exhaustive, and new research is continually being published in this domain, now more than ever. We realize that the significance of our research is somewhat constrained by our paper selection criteria since we were only able to evaluate works published in peer-reviewed publications and were written in English for this study. Furthermore, since our study provides detailed descriptions of the various specifications of the papers, we reviewed, its ability to broadly investigate this domain was limited.

We have included an overview of the current challenges and future implications in the domain of ML applications in various healthcare activities using EHR data in Sect. 4 of this article which we believe will help in promoting future research and administrative decision making.

To decrease the potential of bias in the collection of data, three researchers systematically assessed and identified potential articles, with disagreements addressed by full-text examination and discussion. Each article was methodically elicited important information, and a common data extraction sheet was used to store such information systematically. Reports and interpretations were constantly examined and re-evaluated by all members of the study to make sure that the descriptions of diverse techniques and findings were balanced.

References

Abdel-Basset, M., Chang, V., Hawash, H., Chakrabortty, R. K., & Ryan, M. (2021). FSS-2019-nCov: A deep learning architecture for semi-supervised few-shot segmentation of COVID-19 infection. *Knowledge-Based Systems, 212*, 106647. https://doi.org/10.1016/j.knosys.2020.106647

Alankar, B., Kaur, H., Ahsaan, S., Sharma, G., & Chang, V. (2021). Towards Reviewing an Immediate Impact of COVID-19 on the Integrative World Economy: An Evolving Perspective. *Journal of Global Information Management (JGIM), 30*(4), 1–19. https://doi.org/10.4018/JGIM.20220701.oa6

Banerjee, A., Ray, S., Vorselaars, B., Kitson, J., Mamalakis, M., Weeks, S., & Mackenzie, L. S. (2020). Use of machine learning and artificial intelligence to predict SARS-CoV-2 infection from full blood counts in a population. *International Immunopharmacology, 86*, 106705. https://doi.org/10.1016/j.intimp.2020.106705

Banoo, S., Bell, D., Bossuyt, P., Herring, A., Mabey, D., Poole, F., & Ramsay, A. (2007). Evaluation of diagnostic tests for infectious diseases: General principles. *Nature Reviews Microbiology, 5*(11), S21–S31. https://doi.org/10.1038/nrmicro1523

Brinati, D., Campagner, A., Ferrari, D., Locatelli, M., Banfi, G., & Cabitza, F. (2020). Detection of COVID-19 infection from routine blood exams with machine learning: A feasibility study. *Journal of Medical Systems, 44*(8), 1–12. https://doi.org/10.1007/s10916-020-01597-4

CoVID-19 Community Mortality Risk Prediction (CoCoMoRP) Tool: https://ashis-das.shinyapps.io/CoCoMoRP/

Dagliati, A., Malovini, A., Tibollo, V., & Bellazzi, R. (2021). Health informatics and EHR to support clinical research in the COVID-19 pandemic: An overview. *Briefings in Bioinformatics, 22*(2), 812–822. https://doi.org/10.1093/bib/bbaa418

D'Ambrosia, C., Christensen, H., & Aronoff-Spencer, E. (2020). Computing SARS-CoV-2 infection risk from symptoms, imaging, and test data: Diagnostic model development. *Journal of Medical Internet Research, 22*(12), e24478. https://doi.org/10.2196/24478

Das, A. K., Mishra, S., & Gopalan, S. S. (2020). Predicting CoVID-19 community mortality risk using machine learning and development of an online prognostic tool. *PeerJ, 8*, e10083. https://doi.org/10.7717/peerj.10083

Davenport, T., & Kalakota, R. (2019). The potential for artificial intelligence in healthcare. *Future Healthcare Journal, 6*(2), 94. https://doi.org/10.7861/futurehosp.6-2-94

Dhiman, G., Chang, V., Kant Singh, K., & Shankar, A. (2021). Adopt: automatic deep learning and optimization-based approach for detection of novel coronavirus covid-19 disease using x-ray images. *Journal of biomolecular structure and dynamics*, 1-13. doi: https://doi.org/10.1080/07391102.2021.1875049

Fernandes, F. T., de Oliveira, T. A., Teixeira, C. E., de Moraes Batista, A. F., Dalla Costa, G., & Chiavegatto Filho, A. D. P. (2021). A multipurpose machine learning approach to predict COVID-19 negative prognosis in São Paulo. *Brazil. Scientific Reports, 11*(1), 1–7. https://doi.org/10.1038/s41598-021-82885-y

Fink, A. (1998). *Conducting literature research reviews: From paper to the Internet*. Sage Publications Inc.

Gao, Y., Cai, G. Y., Fang, W., Li, H. Y., Wang, S. Y., Chen, L., & Gao, Q. L. (2020). Machine learning based early warning system enables accurate mortality risk prediction for COVID-19. *Nature Communications, 11*(1), 1–10. https://doi.org/10.1038/s41467-020-18684-2

GitHub: Metadata of Coronavirus https://github.com/Akibkhanday/Meta-data-of-Coronavirus

Goodman-Meza, D., Rudas, A., Chiang, J. N., Adamson, P. C., Ebinger, J., Sun, N., & Manuel, V. (2020). A machine learning algorithm to increase COVID-19 inpatient diagnostic capacity. *PLoS ONE, 15*(9), e0239474. https://doi.org/10.1371/journal.pone.0239474

Gupta, A., Madhavan, M. V., Poterucha, T. J., DeFilippis, E. M., Hennessey, J. A., Redfors, B., & Parikh, S. A. (2021). Association between antecedent statin use and decreased mortality in hospitalized patients with COVID-19. *Nature Communications, 12*(1), 1–9. https://doi.org/10.1038/s41467-021-21553-1

Hsieh, K. L., Wang, Y., Chen, L., Zhao, Z., Savitz, S., Jiang, X., & Kim, Y. (2020). Drug repurposing for covid-19 using graph neural network with genetic, mechanistic, and epidemiological validation. *Research Square*

JHU: John Hopkins University(2020). COVID-19 Dashboard by the Center for Systems Science and Engineering (CSSE) at Johns Hopkins University (JHU). https://coronavirus.jhu.edu/map.html (accessed July 27, 2021).

Joseph, P. C., Paul, M., Lan, D. (2020) *Open database of COVID-19 cases with chest X-ray or CT images* . Git Hub. https://github.com/ieee8023/covid-chestxray-dataset

Joshi, R. P., Pejaver, V., Hammarlund, N. E., Sung, H., Lee, S. K., Furmanchuk, A. O., & Banaei, N. (2020). A predictive tool for identification of SARS-CoV-2 PCR-negative emergency department patients using routine test results. *Journal of Clinical Virology, 129*, 104502. https://doi.org/10.1016/j.jcv.2020.104502

Karthikeyan, A., Garg, A., Vinod, P. K., & Priyakumar, U. D. (2021). Machine learning based clinical decision support system for early COVID-19 mortality prediction. *Frontiers in public health, 9*

Khanday, A. M. U. D., Rabani, S. T., Khan, Q. R., Rouf, N., & Din, M. M. U. (2020). Machine learning based approaches for detecting COVID-19 using clinical text data. *International Journal of Information Technology, 12*(3), 731–739. https://doi.org/10.1007/s41870-020-00495-9

Kim, E., Rubinstein, S. M., Nead, K. T., Wojcieszynski, A. P., Gabriel, P. E., & Warner, J. L. (2019, October). The evolving use of electronic health records (EHR) for research. In *Seminars in radiation oncology* (Vol. 29, No. 4, pp. 354–361). WB Saunders. doi: https://doi.org/10.1016/j.semradonc.2019.05.010

Kukar, M., Gunčar, G., Vovko, T., Podnar, S., Černelč, P., Brvar, M., & Notar, M. (2021). COVID-19 diagnosis by routine blood tests using machine learning. *Scientific Reports, 11*(1), 1–9. https://doi.org/10.1038/s41598-021-90265-9

Langer, T., Favarato, M., Giudici, R., Bassi, G., Garberi, R., Villa, F., & Fumagalli, R. (2020). Development of machine learning models to predict RT-PCR results for severe acute respiratory syndrome coronavirus 2 (SARS-CoV-2) in patients with influenza-like symptoms using only basic clinical data. *Scandinavian Journal of Trauma, Resuscitation and Emergency Medicine, 28*(1), 1–14. https://doi.org/10.1186/s13049-020-00808-8

Liberati, A., Altman, D. G., Tetzlaff, J., Mulrow, C., Gøtzsche, P. C., Ioannidis, J. P., & Moher, D. (2009). The PRISMA statement for reporting systematic reviews and meta-analyses of studies that evaluate health care interventions: Explanation and elaboration. *Journal of Clinical Epidemiology, 62*(10), e1–e34. https://doi.org/10.1371/journal.pmed.1000097

Lotfi, M., Hamblin, M. R., & Rezaei, N. (2020). COVID-19: Transmission, prevention, and potential therapeutic opportunities. *Clinica Chimica Acta, 508*, 254–266. https://doi.org/10.1016/j.cca.2020.05.044

Mahase, E. (2021). Covid-19: What new variants are emerging and how are they being investigated?. doi: https://doi.org/10.1136/bmj.n158

ML-based COVID-19 test from routine blood test Web App: https://covid19-blood-ml.herokuapp.com/

Mohapatra, S., Nath, P., Chatterjee, M., Das, N., Kalita, D., Roy, P., & Satapathi, S. (2020). Repurposing therapeutics for COVID-19: Rapid prediction of commercially available drugs through machine learning and docking. *PLoS ONE, 15*(11), e0241543. https://doi.org/10.1371/journal.pone.0241543

El Naqa, I., & Murphy, M. J. (2015). What is machine learning?. In *machine learning in radiation oncology* (pp. 3–11). Springer, Cham. doi: https://doi.org/10.1007/978-3-319-18305-3_1

Parchure, P., Joshi, H., Dharmarajan, K., Freeman, R., Reich, D. L., Mazumdar, M., & Kia, A. (2020). Development and validation of a machine learning-based prediction model for near-term in-hospital mortality among patients with COVID-19. *BMJ Supportive & Palliative Care*. https://doi.org/10.1136/bmjspcare-2020-002602

Paul, M (2019). *Chest x-ray images (Pneumonia)*. Kaggle. https://www.kaggle.com/paultimothymooney/chest-xray-pneumonia.

Pawlowski, C., Puranik, A., Bandi, H., Venkatakrishnan, A. J., Agarwal, V., Kennedy, R., & Soundararajan, V. (2021). Exploratory analysis of immunization records highlights decreased SARS-CoV-2 rates in individuals with recent non-COVID-19 vaccinations. *Scientific Reports, 11*(1), 1–20. https://doi.org/10.1038/s41598-021-83641-y

Phillips-Wren, G., & Ichalkaranje, N. (Eds.). (2008). *Intelligent decision making: An AI-based approach* (Vol. 97). Springer Science & Business Media. doi: https://doi.org/10.1007/978-3-540-76829-6_.

Pickering, C., & Byrne, J. (2014). The benefits of publishing systematic quantitative literature reviews for PhD candidates and other early-career researchers. *Higher Education Research & Development, 33*(3), 534–548. https://doi.org/10.1080/07294360.2013.841651

Plante, T. B., Blau, A. M., Berg, A. N., Weinberg, A. S., Jun, I. C., Tapson, V. F., & Adib, A. B. (2020). Development and external validation of a machine learning tool to rule out COVID-19 among adults in the emergency department using routine blood tests: A large, multicenter, real-world study. *Journal of Medical Internet Research, 22*(12), e24048. https://doi.org/10.2196/24048

Rechtman, E., Curtin, P., Navarro, E., Nirenberg, S., & Horton, M. K. (2020). Vital signs assessed in initial clinical encounters predict COVID-19 mortality in an NYC hospital system. *Scientific Reports, 10*(1), 1–6. https://doi.org/10.1038/s41598-020-78392-1

2016 Report To Congress on Health IT Progress. Washington (DC): Office of the National Coordinator for Health Information Technology (ONC) Office of the Secretary, United States Department of Health and Human Services, 2016. p 32. https://www.healthit.gov/sites/default/files/ 2016_report_to_congress_on_healthit_progress.pdf . Accessed July 31, 2021.

Ryan, L., Lam, C., Mataraso, S., Allen, A., Green-Saxena, A., Pellegrini, E., & Das, R. (2020). Mortality prediction model for the triage of COVID-19, pneumonia, and mechanically ventilated ICU patients: A retrospective study. *Annals of Medicine and Surgery, 59*, 207–216. https://doi.org/10.1016/j.amsu.2020.09.044

Schöning, V., Liakoni, E., Baumgartner, C., Exadaktylos, A. K., Hautz, W. E., Atkinson, A., & Hammann, F. (2021). Development and validation of a prognostic COVID-19 severity assessment (COSA) score and machine learning models for patient triage at a tertiary hospital. *Journal of Translational Medicine, 19*(1), 1–11. https://doi.org/10.1186/s12967-021-02720-w

Shiri, I., Sorouri, M., Geramifar, P., Nazari, M., Abdollahi, M., Salimi, Y., & Zaidi, H. (2021). Machine learning-based prognostic modeling using clinical data and quantitative radiomic features from chest CT images in COVID-19 patients. *Computers in Biology and Medicine, 132*, 104304. https://doi.org/10.1016/j.compbiomed.2021.104304

Sohrabi, C., Alsafi, Z., O'neill, N., Khan, M., Kerwan, A., Al-Jabir, A., ... & Agha, R. (2020). World Health Organization declares global emergency: A review of the 2019 novel coronavirus (COVID-19). *International journal of surgery, 76*, 71–76. https://doi.org/10.1016/j.ijsu.2020.02.034

Soltan, A. A., Kouchaki, S., Zhu, T., Kiyasseh, D., Taylor, T., Hussain, Z. B., & Clifton, D. (2020). Artificial intelligence driven assessment of routinely collected healthcare data is an effective screening test for COVID-19 in patients presenting to hospital. *MedRxiv*. https://doi.org/10.1016/S2589-7500(20)30274-0

Sutton, R. T., Pincock, D., Baumgart, D. C., Sadowski, D. C., Fedorak, R. N., & Kroeker, K. I. (2020). An overview of clinical decision support systems: Benefits, risks, and strategies for success. *NPJ Digital Medicine, 3*(1), 1–10. https://doi.org/10.1038/s41746-020-0221-y

WHO: World Health Organization, 2020. Coronavirus disease (COVID-19) Weekly Epidemiological Update and Weekly Operational Update https://www.who.int/emergencies/diseases/novel-coronavirus-2019/situation-reports/ (accessed July 27, 2021).

Wolff, R. F., Moons, K. G., Riley, R. D., Whiting, P. F., Westwood, M., Collins, G. S., & Mallett, S. (2019). PROBAST: A tool to assess the risk of bias and applicability of prediction model studies. *Annals of Internal Medicine, 170*(1), 51–58. https://doi.org/10.7326/M18-1376

Yadaw, A. S., Li, Y. C., Bose, S., Iyengar, R., Bunyavanich, S., & Pandey, G. (2020). Clinical features of COVID-19 mortality: Development and validation of a clinical prediction model. *The Lancet Digital Health, 2*(10), e516–e525. https://doi.org/10.1016/S2589-7500(20)30217-X

Machine Learning-Based Emerging Technologies in the Post Pandemic Scenario

R. L. Priya, A. Abirami, and Naman Desai

Abstract The entire world is now fighting to overcome the after-effects of the COVID-19 pandemic. The most challenging factor is handling the post-pandemic scenario in various sectors such as medical, educational, civil services, business, etc. Every sector put its efforts in the best possible way to cope up with these unprecedented times. The majority of the sectors came up with the idea of work from home, robotics, online payment system, entertainment, meetings, webinars, etc., to continue with the smooth working environment. To manage this crisis, it is essential to be equipped with appropriate tools and technologies. These technologies may differ from sector to sector to satisfy their essential requirements. Among these, the educational sector is the only domain in which the entire transition happened from the classroom model to a fully virtual model of service. The digitalization of the educational system ensured the continuous delivery of education without any requirement of physical presence. The proposed idea uses data science techniques to understand the crisis in a better way and improvise the quality of education. In the medical sector, there are various techniques like online consultation with doctors, telemedicine and chatbots. Augmented Reality solutions provide a better platform for convenient, smooth and safe doctor-patient interaction. Data science techniques are implied to build an efficient novel approach for early diagnosis in medical sectors and reduce the mortality rate. Similarly, in other sectors such as business and civil services, there are betterments going to be brought about in a virtual environment. The segment which is facing the repercussions of this deadly virus is the common man, especially farmers, small businessmen, daily wagers, elders, children, and patients. Considering the current situation of the common public, one should try to make a

R. L. Priya (✉)
Department of Computer Engineering, VES Institute of Technology Mumbai, Mumbai, India
e-mail: priya.rl@ves.ac.in

A. Abirami
Department of Information Technology, Bannari Amman Institute of Technology Erode, Erode, Tamilnadu, India

N. Desai
Department of Information Technology, Shah and Anchor Kutchhi Engineering College, Mumbai, India

V. Chang et al. (eds.), *Artificial Intelligence and Machine Learning Methods in COVID-19 and Related Health Diseases*, Studies in Computational Intelligence 1023,
https://doi.org/10.1007/978-3-031-04597-4_3

brave new world, which is more adaptable, sustainable and stronger to mitigate the risks. This system includes a CNN model that classifies if a pair of lungs is normal or has pneumonia. The CNN model provides 87.71% accuracy, 12.29% miscalculation rate, 80.45% precision, 80.94% sensitivity, 91.20% specificity, 80.03% F-1 value against the testing dataset having a loss value of 0.5096.

Keywords Data science · Covid-19 · Post-pandemic · Artificial intelligence · Machine learning · Deep learning and neural networks

1 Introduction

The World Health Organization (WHO) declared Covid-19 as a pandemic in March 2020 (Rathore et al., 2020), after which the entire world has come to a pause as the lockdown was implemented to curb the spread of the virus. Various sectors in the world have been deeply affected. This pandemic hit the United States of America severely, leading to a widespread social distancing and closure of workplaces too. Densely populated and developing countries like India are incapable of reducing the spread of this virus and hence have reported a huge number of patients. Efforts have been taken by the national as well as the state government of India and many other countries to bring this pandemic under control. Health check posts have been established to monitor people at their arrival from other countries. Hospitals in countries massively affected by pandemics are experiencing an increase in cost and decrease in revenues, thus threatening their financial stability. Covid-19 had already caused significant loss of life in many countries around the globe. Nasopharyngeal swab or broncho-alveolar lavage polymerase chain reaction rapid testing for the SARS-CoV-2 genetic material are the two tests done in larger numbers for detecting the virus. Another means to access COVID-19 status can be antibody testing. It provides a clearer understanding of the occurrence and fatality rate of the disease as it could easily determine the patients are having mild or even no symptoms. As SARS-CoV-2 is a newly identified pathogen, there is no pre-existing immunity or cure to it. Elderly people with chronic issues and weak immune systems are comparatively more vulnerable to the virus. According to reports from China, people above 70 years have a fatality rate of 80% (Wang et al., 2020). It was observed that people having hypertension, diabetes, cardiovascular disease have significantly higher risk factors.

In many countries, strong and collaborative relationships have been built with the state and other regional authorities. It had proved to be useful for coordination and to take action quickly during emergencies. Many innovative solutions have been proposed in order to tackle this ongoing pandemic. Culture cannot be changed without the support of citizens and hence they play a crucial role in building and maintaining relationships among nations. Local leaders should understand the plight of the citizens regarding health services and should work together to find practical solutions in order to ease their plight. Due to the uncertain pattern of Covid-19, many orders and norms proposed by the states have been discarded. Nursing communities across the world have been going through unforeseen challenges in this pandemic (Hořínka & Kontriková, 2020). Moreover, the spread of Covid-19 has brought many restrictions to the travel sector. People are advised not to travel via trains or any other public transport amidst these difficult times. It is mandatory for everyone to practice social distancing i.e., reducing interactions between people by maintaining a particular distance from them. It might completely change the outdoor activities performed by people. The demand for travel will considerably reduce after this pandemic. Social distancing can negatively affect health status and social well-being in society. A number of countries have implemented various strategies to avoid contact as much as possible. The spread of covid-19 is majorly due to respiratory droplets and being in close proximity to each other. It is uncertain until now how long social distancing will remain an absolute necessity.

On the positive side, as travel and mobility have reduced considerably, many countries have witnessed an unprecedented drop in car traffic, thus decreasing air and noise pollution. Although it is termed as a temporary solution as in the future the demand for travel is certain to rise, people might fear social contact even after social distancing rules are lifted up post to the advent of the vaccine and due to practices like work from home and e-learning, people are more likely to carry out activities at home, with their family rather than traveling hence leading to less traffic and congestion during office hours. Another upside is the rise in online delivery options. These options have made people more likely to stay at home instead of moving out for shopping.

2 Risk Factors during COVID-19

(CDC—Centre for Disease Control and Prevention)

COVID-19 can affect any person and the disease causes from mild to very severe symptoms. For other illnesses caused by respiratory viruses (such as influenza), some people may be more likely to have severe illnesses than others because they have medical conditions that increase their risk. These are commonly termed as "risk factors". Examples include older age or having certain underlying medical conditions (Tang, 2020).

CDC is conducting disease surveillance and field investigations to understand why some people are more likely to develop severe COVID-19 illness. This has become the top priority in CDC's strategy to combat COVID-19. We learn from these efforts to provide vital information to help CDC scientists and other public health officials make decisions to protect the most vulnerable populations.

To maximize their output, CDC carries out these activities in collaboration with state, local, and territorial health departments, public health, commercial, clinical laboratories, vital statistics offices, health care providers, emergency departments, academic and private sector partners.

2.1 *Why Risk Factors Matter?*

People with risk factors may be more likely to need hospitalization and intensive care if they have COVID-19 else, they may be more likely to succumb to the infection.

It is essential to understand the risk factors for severe COVID-19 illness, which are to:

(1) Take precautions as you go about your daily life and attend events.
(2) Better understand how a medical condition could affect your own health if you get sick with COVID-19.
(3) Anticipate medical treatment that you might need if you get sick.
(4) Reduce your risk for severe COVID-19 illness by managing any conditions you have that are risk factors.

2.2 *How we learn about Risk Factors for severe disease?*

Because COVID-19 is a newly mutated disease, a more comprehensive and in-depth study is needed to understand the risk factors better for severe illness or dire complications. The potential risk factors that have been identified include:

(1) Age.
(2) Race/ethnicity.
(3) Gender.
(4) Some medical conditions.
(5) Use of certain medications.
(6) Poverty and crowding.
(7) Certain occupations.
(8) Pregnancy.

Additional research will help confirm if these are risk factors for severe COVID-19 illness and whether or not other factors increase a person's risk.

3 Impacts of COVID-19 in major sectors

The major sectors under the government are categorized in Fig. 1. Over 7.5 million people have been infected by COVID-19 globally (Udupa et al., 2020). In India, the number of cases rose exponentially from 470 in March to a staggering number of 4 lakhs within three months. Despite the implementation of lockdown, the cases were increasing exponentially. The number continues to rise as the lockdown measures are being relaxed in varying capacities across countries.

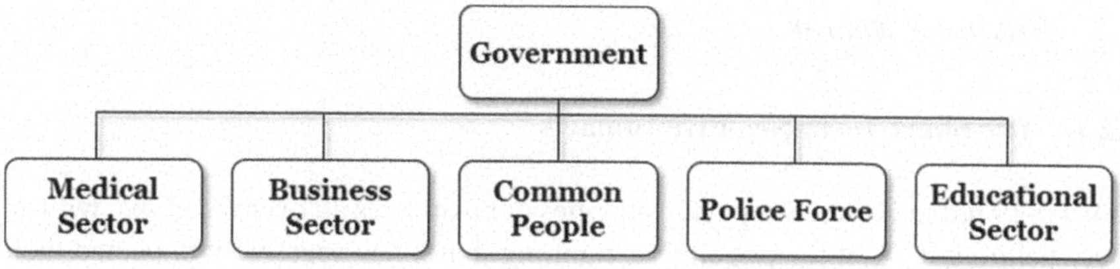

Fig. 1 Major sectors under the government

3.1 Medical Sector

A huge number of policy measures have been implemented to contain the spread of COVID-19. It resulted in significant operational disruption for all types of welfare companies. Staff quarantine, supply-chain failures, and sudden increments in customer demand have generated serious complications for welfare companies across a wider range of sectors than anticipated. The revenue that is lost in this period is permanent loss and has and will continue to put sudden and unanticipated pressure on working capital and the liquidity ratio.

Rising positive COVID-19 cases are also generating the demand for medical ventilators across the world. It has been observed that around 5% of the total positive COVID-19 cases are severe that require ventilators to ensure proper breathing. This resulted in an exponentially growing demand for medical ventilators across the globe. Most of the welfare companies have extended their production capacity to meet the ongoing demand. For instance,

(1) Getinge companies have increased their capacity by 60% compared to their previous production capacity of 10,000.
(2) The company Philips has doubled its ventilator production capacity from ~ 1000 ventilators a week to ~ 2000 ventilators.
(3) The Medtronic Company announced to upscale its production by 40% and is working hard to double it in the coming days.
(4) GE Healthcare companies have come up with an agreement with Ford to fill the gap between demand and supply (Nabi & Mishra, 2021).

It has been observed that many companies are revamping their production criterion to combat COVID-19. There are many industries and luxury hotels that are entering into the masks and quarantine center services. Many new start-ups are also contributing to the fight against COVID-19 by introducing apps and other features that help track COVID-19. Apart from a rise in demand and a shortage of medical supply capacities, the spread of medical protection has been one of the biggest problems for supply chains in healthcare yet.

3.2 *Business Sector*

3.2.1 The Short Term Negative Impacts

COVID-19 has negatively impacted all types of businesses all over the globe, with an overwhelming 93% of the population claiming that it has negatively impacted their business.

Specifically, only 28% of the entrepreneurs expect a long-term negative impact, while 65% expect a short-term negative impact on their business. It goes without saying that the negative impact will depend on the nature of the business, but the most obvious one would be on the revenues of the company (Gopisetty, 2020).

There is an expected rise in companies having better economic conditions from 25 to 51% from March to October of 2020 and a corresponding decrease in companies having worse economic conditions from 58 to 29% from March to October of 2020 as compared to the current economic conditions prior to the pandemic.

3.2.2 Revenue Falls on Businesses

Businesses have been taking major hits since the beginning of 2019 due to the rapid spread of COVID-19. Our survey results show that 86% of the entrepreneurs are expecting their revenue to take significant falls due to the direct impact of the pandemic.

From the 86% of the entrepreneurs, 33% of the entrepreneurs are anticipating a revenue fall of more than 15% in their business, 21% believe their business' revenue will fall between 10 to 15%, while another 33% of the entrepreneurs expects their revenue to fall 10% or less.

The significant drop in revenue will negatively affect the economy as companies are struggling to adjust their business strategies and begin preparations for an economic downturn (Gopisetty, 2020).

The Economic Downturn

From closing down productions and operations to a massive drop to consumer confidence, COVID-19 continues to impact businesses.

77% of the entrepreneurs believe that an economic downturn is inevitable as the government will have to reinstate more restrictions and the continued threat of health to consumers would become a hazard.

Countries like Italy, China, and India are expected by prediction to take the greatest personal financial impact (Gopisetty, 2020).

3.2.3 Business Bounce Back

Despite the current situation, many of the entrepreneurs are optimistic and confident in their businesses' ability to bounce back and have their operations and economic earning a return to normal after the government restrictions will have been lifted.

Only 12% of the entrepreneurs believe that their operations and economics will take six months to over a year to return to optimal or general levels. This is a huge gap for the 88% of the entrepreneurs who expect to have their businesses back to operational levels and sustainable developments in less than six months.

From that 88% of the entrepreneurs, 32% of the entrepreneurs are expected to be operational in less than three days, while another 32% of the entrepreneurs expect daily operations to return to normal in 60 to 90 days. Lastly, 24% of the entrepreneurs believe their organizations and companies will be back to normal in at least half a year (Gopisetty, 2020).

Shifting to Remote Areas of Work

Due to government restrictions and the shutdown of non-essential services and hence businesses, many companies and businesses have opted for remote working to ensure that COVID-19 does not spread within their organization to continue their productions.

73% of the entrepreneurs have decided to shift their business into a fully-remote working environment to continue their business operations. In addition to remote working, 67% of the entrepreneurs have also decided to postpone business events and activities to reduce social contact and spread the COVID-19 virus (Gopisetty, 2020).

Common People

The pandemic has affected every individual and laid bare their fragility. Border closures, trade restrictions and confinement measures have been and will continue to be a preventative measure for farmers and agriculturists from accessing markets, including for buying inputs and selling their produce and farmers from harvesting their various crops, thus disrupting not only domestic but also international dietary supply chains and reducing access to healthy, safe and diverse food habits. The pandemic has decimated jobs and placed millions of livelihoods on a thread. As breadwinners (daily wages) lose jobs, fall ill and die, the food security and nutrition of millions of individuals is under threat, with those in low-income, developing countries, particularly the most marginalized populations per country, which include small-scale farmers and indigenous and simple people, being hit the hardest.

Billions of agricultural workers, waged and self-employed individuals while feeding the world, regularly face high levels of working poverty, malnutrition and poor health and suffer from a lack of safety and labor protection as well as other types of abuse, which is illegitimate. With very low and highly irregular incomes and a moral lack of social support, many of them are spurred to continue working, often in very unsafe conditions, thus exposing themselves and their families to high additional risks. Further, when experiencing mass income losses, they may resort to negative coping marketing strategies, such as distress sale of assets, predatory loans, or even child labor. Migrant agricultural workers are particularly highly vulnerable because they face horrid risks in their transport, working and living conditions and struggle every day just to access their support measures put in place by governments. Guaranteeing the safety and health of all agriculturists from primary producers to those involved in the depths of food processing, transport and retail, including street food vendors and as well as better incomes and protection, will be critical to saving lives and protecting the health of the public.

In the COVID-19 crisis, food security, public health, employment and labor issues, particularly individuals' health and safety, are in jeopardy. Adhering to workplace safety and health practices will ensure access to decent work and the protection of labor rights in all industries and this will be crucial in addressing the human dimension of the pandemic. Immediate action to save lives and livelihoods would include extending social protection towards universal health coverage and monetary income support for those most affected by the pandemic. These include workers in the informal economy and in very poorly protected and low-paid jobs, including youth, older workers and migrant individuals. Particular attention must be paid to the situation of women, who are now over-represented in low-paid jobs and care role jobs. Different forms of support should be given like cash transfers, children allowances and healthy school meals, shelter and food relief initiatives.

3.3 Police Force

Policing is one of the most disturbing activities in the world as opposed to long and frequent changes, threats of violence, the growing need for self-control and the lack of social support that cause great stress. As a result, law enforcement officials experience mental health problems to a much greater degree than most people, even before they face the challenges of epidemics, depression and uncertainty. Mental health problems among legal practitioners are associated with the workplace, agency culture, consistent planning, and exposure to potentially high levels of traumatic events and post-traumatic stress disorder. As a result, law enforcement officials are more likely to suffer from depression, family conflicts, alcohol abuse, and suicide attempts. These negative behaviors can prevent the officer's resilience after traumatic events. They also have the power to disable responses to major social and occupational changes caused by unfavorable conditions that place the growing demand for various law enforcement resources.

3.3.1 COVID-19 Related Operational Measures

Eighty-three percent of police officers have realized they have reduced public access to their facilities. The same percentage reduced their use of community policing services. Accordingly, public policing has been reduced by more than 80%.

As crime is often unreported—as it shows a significant difference between reported crime and research of crime victims—the importance of community policing to close this gap is significant. Contrary to what has happened in the past, access to police departments has decreased so that physical availability and public connections to the police have been reduced.

Suspension of civilian and academic rides and the cancellation of community forums by the police could disrupt police and civilian solidarity. Using virtual forums, conference calls and access to community leaders in the non-physical world can prevent road level restrictions, (physical) social links caused by the epidemic.

Reduced enforcement actions and personal responses to service calls were found between 78 and 76% of respondents, respectively. Collectively, this decline raises questions about how LE can meet the needs of society at the same level as before COVID-19. Obstacles to traditional policing imposed by COVID-19 have to be dealt with until the spread of the disease has subsided.

New types of enforcement actions involving face-to-face, local shelter orders, and newly appointed police are important new functions of the police. Law enforcement is placed on the unequal role of enforcing epidemic-related laws as opposed by other states with a voice. Interestingly, while face masks are encouraged to reduce the transmission of the virus, criminals have been used to conceal their identity.

Accordingly, arrests, vehicle stops and investigations have dropped dramatically nationwide with fewer vehicles on the roads. Shootings, killings, domestic violence and robberies have been reported in some major cities during the epidemic.

The ongoing segregation of parts of US society has also led to various criminal behaviors: hate crime, carjacking, arson, and vandalism of public and private property by people of all walks of life. Law enforcement must also deal with these new oppressors (Stogner et al., 2020).

3.3.2 COVID-19 Related Training Measures

The survey found that 82% of police agents set up their own police schools and vocational training camps.

Stopping the flow of employees delays the transition of those who are retired, resigned, or ill. The need to introduce improved police training, especially in view of the challenges posed by the epidemic and protests, is still paramount.

Similarly, 80% of departments have changed the roll call process. That the changes in telecommunications are hindering the sharing of information, the merger of departments and the effectiveness of the policy remains to be seen (Stogner et al., 2020).

3.3.3 COVID-19 Related Staffing Measures

68% of agencies changed their staff planning. Depending on the circumstances, the officers were assigned to work remotely and separately. Once again, what impact will this adjustment have on productivity, efficiency, collaboration and information sharing within departments during the epidemic?

Sixty-two percent of departments indicated that they were reducing staff access to certain police stations. What are the implications for police performance from this reduction of physical intimacy between colleagues?

As elsewhere in the US labor market, 30% of police departments said they were overcrowded or understaffed because of COVID-19. These pressures on employees can result in staff shortages and declining morals.

Even before the epidemic, law enforcement officials were having difficulty recruiting new officers and subsequent appointments. Budgetary pressures resulting from a reduction in state and domestic revenues due to declining business activity during the epidemic will pose additional challenges to funding new entrants to police departments. Also, the emergence of calls for overpayment of the police or a reduction in other police budgets going forward reveals unprecedented financial uncertainty and adverse effects on the entire employment and personnel environment (Stogner et al., 2020).

3.4 *Educational Sector*

Attending school is the best social policy tool available to develop social skills. While school time can be fun and can enhance social skills and public awareness, from an economic point of view, the main point of being in school is that it increases a child's ability. Even a short time in school we do this; even a shorter period of time compared to a lost school will have consequences for skills development. But can we estimate how much of a COVID-19 disruption will affect learning? Not necessarily, since we are in the new world, but we can use other courses to get a larger order.

The closure of schools, colleges, and universities disrupts student education around the world and leads to the postponement of many exams. Internal testing is probably considered irrelevant and most have simply been cancelled. But its purpose is to provide a graph of a child's development to families and teachers. Loss of this information delays the recognition of great difficulty and learning difficulties and can have long-term negative effects on the child.

Here is a summary of all the changes Covid-19 has included in the Education Program.

March 2020 will be remembered forever in history as a time when all schools, colleges and educational institutions around the world close. Although no one was ready for the global epidemic, the education system was one of the most affected sectors.

According to a United Nations report, the closure of schools and other educational institutions has affected 94% of the world's students. It has affected nearly 1.6 billion students in more than 190 countries and across continents.

However, it is commendable how the international education sector has adapted quickly and efficiently and consistently. On the other hand, this crisis has revived some unprecedented discoveries in the field of education and has given all of us a new appreciation for the role of teachers and educators. On the other hand, the disease has highlighted the extreme economic and social inequalities among students. We have seen the need to better train teachers in new ways of delivering education. The global impact of Covid-19 is multifaceted and is evident in almost all sectors, especially the health, economic and educational sectors. Since the virus was announced as a pandemic in March 2020, there have been dozens of daily reports of its impact on the lives of millions worldwide. Similarly, a major national concern has been to reduce the spread of the virus and reduce its effects on society as a whole and high-risk communities (Khanna & Prasad, 2020).

More than 1.5 billion students in some 165 countries have been affected by the closure of schools and campuses. As a result, schools, colleges and universities are being forced to make some changes in online learning as a substitute for site delivery (Khanna & Prasad, 2020).

3.4.1 Graduates

Jobs for university graduates this year could be severely affected by the COVID-19 epidemic. They once faced major educational interruptions in the latter part of their studies, experienced major disruptions in their examinations, and eventually were likely to succeed at the start of a major global economy. Evidence suggests that poor market conditions in the labor market cause workers to accept low-paying jobs and that this has long-term effects on the jobs of some.

We can see below in Table 1 how the education sector of the various countries was affected by COVID–19 between the 16th of February and 21st of November of 2020.

4 Data Science methodologies to overcome impacts in major sectors

During the course of the COVID-19 pandemic, the world has witnessed rapid advancements in technology pertinent to the department of healthcare. The fast-paced technological development that has led to an increase in computational power and storage capacities has carved a niche for emerging technologies like Edge Computing, Deep Learning, Computer Vision, and the Internet of Things. The innovations during the pandemic in Medicine and Health are truly unprecedented and harness the power

Table 1 Global plots and statistics (Khanna & Prasad, 2020)

Date	Affected learners	% of total enrolled learners	Country-wide closures	Countries partially open	Countries closed	Countries fully open	Countries having an academic break
16/02/2020	999,014	0.1	1	China	N.A	India, UK, Australia, Canada, USA, etcetera	NA
16/06/2020	945,229,514	54	96	China, UK, USA, etcetera	India, Canada, etcetera	Australia, Sweden, etcetera	Kazakhstan, Sudan, Finland, Pakistan, etcetera
16/10/2020	265,640,265	15.2	26	Canada, USA, India, etcetera	Brazil, Mexico, etcetera	Russia, China, Australia, etcetera	South Africa, Sudan, Senegal, Oman, etcetera
21/11/2020	224,068,338	12.8	23	Canada, USA, India, etcetera	Brazil, Mexico, etcetera	Russia, China, Australia, etcetera	Sudan

of Data Science, Computer Vision, and Image Processing Techniques to counter the shortcomings of providing enough resources to the entire population suffering from the virus. The methodologies to overcome impacts of COVID-19 in major sectors under the government can be seen in Fig. 2.

4.1 Medical Sector

With the advent of faster computational power and stellar GPU systems, autonomous systems have truly been the buzz of the twenty-first century. From embedding Artificial Intelligence in bots, making electric vehicles to data analysis with the Internet of Things and Edge Computing, this decade has seen a revolution unlike ever before. Although a myriad of health experts compare the current pandemic situation to a 'Bio-War', the emergence of high-tech solutions to combat this pandemic has become the need of the day. Social distancing and improvised health facilities to meticulously monitor patient needs would become a norm post-pandemic.

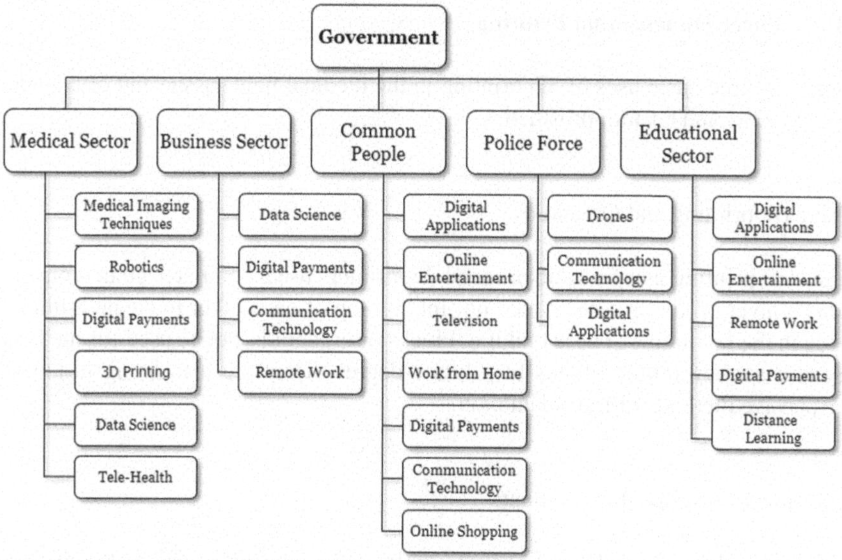

Fig. 2 Methodologies to overcome impacts of COVID-19 in major sectors under the government

4.1.1 Medical Imaging Techniques

To overcome the problem of early screening of symptoms, Chest X-Rays and CT-Scans of the patients are analysed and fed into a Deep Neural Network model which then classifies which respiratory condition the patient is suffering from. These ailments range from detecting forms of Pneumonia, SARS, MERS, COVID-19, ARDS, and many more. Heat maps and Saliency maps could display and highlight the infected part of the patient suffering from the virus. It pinpoints exactly where the infection is located in the lungs. This eliminates the problem of trust.

4.1.2 Data Science by Analysis of Genomic Sequence Analysis

Bioinformatics is another field that has improvised its research owing to data science. The genomic Sequence of the COVID-19 virus is studied using its Protein Sequence. Different amino acids that contribute to the genome of the virus are formulated from the protein sequence and a 3-D molecular structure is synthesized. This tremendously accelerates the progress in the quest of finding a vaccine.

4.1.3 Three Dimensional Printing

There are three basic uses of 3D printing in the medical field: tissue and organoids, surgical tools and custom prosthetics.

Tissue Bioprinting and Organoids

Instead of printing using plastic or metal, bioprinters use a computer-guided pipette to lay a layer on living cells, called bio-ink, on top of each other to create artificial tissue in the laboratory. These building blocks or organoids can be used for medical research as they mimic organs to a lesser degree. They are also being tested as inexpensive methods of transplantation.

3D Printing of Surgical Instruments

3D printing does not only make sterile tools, and some are based on the ancient Japanese tradition of origami, which means they are accurate and can be made much smaller. These devices can be used to operate in small areas without causing unnecessary injury to the patient.

Custom Prosthetics Use 3D Printing

The low price point of these products makes them especially useful for use with children, who are rapidly increasing their artificial limbs. 3D printing also allows the patient to design fittings tailored to their needs.

4.1.4 Digital Payments

Social distancing and minimizing manual contact are paramount when it comes to mitigating the spread of the virus. Generally, hospital bills have to come in contact with a lot of intermediates before they are finally entered into the digital database of the hospital/clinic via the receptionist. In a place like a hospital, where germs are proliferated every now and then, there needs to be a robust method that avoids intermediate contacts altogether. The digitization of bills could be done by using Optical Character Recognition (OCR) to scan the bill, extract the critical sections of the text, and directly enter the main database. At the same time, the patient would get a notification and they could pay their bills through their digital payment systems on their mobile phones.

4.1.5 Robotics

Robots that specialize in human therapies include the use of robots and regenerative robots. The range of auxiliary and medical robotics devices is also growing rapidly. These include robots that help patients recover from trauma such as stroke, take care of the elderly or physically / mentally disabled and industrial robots that perform various routine tasks, such as disinfection rooms and delivery of medical supplies and equipment, including medicines.

Below are the top six uses of robots in the medical field today (Pott et al., 2004).

Telepresence Doctors

These robots help them diagnose and treat patients in rural or remote areas, giving them "telepresence" in the room.

Surgical Assistants

These remote-controlled robots assist surgeons with performing operations, slow-moving procedures.

Rehabilitation Robots

These play an important role in rehabilitating people with disabilities, including improved mobility, energy, coordination, and quality of life. These robots can be programmed to adapt to each patient's condition as he recovers from a stroke, brain injury or spinal cord injury, or neurobehavioral or neuromuscular diseases such as multiple sclerosis. Reality combined with renewable energy robots can improve balance, mobility, and other motor functions.

Robotic Transportation

The Robotic Supplies, medicines, and food are delivered to patients and staff through these robots, thereby enhancing communication between doctors, hospital staff, and patients.

Hygiene and Disinfection Robots

With the growing number of antimicrobials and the spread of deadly diseases such as Ebola, many health care facilities are using robots to clean and disinfect their existing areas.

Robotic Prescription Dispensing Systems

The greatest advantages of robots are speed and accuracy, two of the most important factors in pharmacies.

4.1.6 Telehealth

While the world has slowly started to commence with manufacturing to appease its deteriorating financial situation by lifting some of the restrictions, the lives of the millions of humans serving the community by delivering them the essential goods are still an enormous risk. Even E-Commerce giants like Amazon have fallen prey to the grueling pandemic situation. An excerpt from an interview with an Amazon warehouse worker said: i feel that there's no effort, and they're not taking it seriously. They kept saying they were doing extensive cleaning, extensive cleaning—they weren't. There aren't enough wipes and spray; one wipe is used to clean four people's stations. I said something to my manager, and he just shook his head and said, "It's better than nothing." According to business-standard.com, more than 2.4 million Indians die of treatable conditions every year due to lack of access to medications. Another study revealed that urban cities are suffering from severe pollution and congestion as delivery vehicles choke up the street.

The project undertaken is an attempt to channel cutting edge technologies like Computer Vision, Aerial Robotics, Image Processing and Artificial Intelligence to combat the loopholes in the traditional logistic and supply chain systems that involve extensive human intervention and are much slower while causing deterring effects on the environment The system developed by our team uses Drones to facilitate cost-effective, less time-consuming delivery to even the most rural parts of the country to uplift those devoid of essential services. The USP of our system that differentiates it from the rest is that it autonomously delivers goods directly from the warehouse to the destination with minimal manual control. Other drone-delivery systems are not intelligently programmed to identify the product in the warehouse.

They are only capable of delivery. Once the drone gets the notification in our system, it makes its way towards the warehouse and uses object detection algorithms to find the required product. Then by using Image Processing mechanisms like Adaptive Binary Thresholding and Optical Character Recognition, it scans the QR Code and recognizes the version of the product that the user has requested. Eventually, it finds the most optimal path to the destination using shortest path algorithms like Dijkstra's Algorithm. At the destination, an Aruco Marker would be placed for its Precision Landing. Obstacle Avoidance and Routing Algorithms are internally programmed to optimize its path. Drone Delivery would absolutely be the next big innovation post-pandemic.

4.2 Business Sector

4.2.1 Data Science Methodologies

Most data scientists in the industry are advanced and trained in mathematics, statistics and computer science. Their experience is a great light that also reaches out to data viewing, data mining, and information management. It is common for them to have previous experience in infrastructure construction, cloud computing, and data storage.

Benefits of Data Science in Business

(1) Reduce risk and fraud: Data scientists are trained to identify specific details in a certain way. They create statistics, networks, methods and large data sets for fraudulent trend models and use those to create alerts that help ensure timely responses when unfamiliar data are known.
(2) Delivering the right products: One of the benefits of data science is that organizations can determine when and where their products sell. This can help deliver the right products at the right time, which can help companies create new products to meet the needs of their customers.
(3) Customized customer experience: One of the biggest and most helpful benefits of data science is the ability of sales and marketing teams to understand their audience at a very granular level. With this information, an organization can create the best customer experience.

Ways Data Science Adds Value to Businesses

The points are given below discuss the values that data science promotes:

(1) Empowering Managers and Officials to Make Better Decisions: An experienced scientist may be a trusted advisor and strategic partner in the top management of the organization by ensuring that employees develop their analytical skills. The data scientist deduces and demonstrates the importance of data to make advanced decision-making processes across the organization by measuring, tracking, and recording performance information.
(2) Directing Actions Based on Trends: A data scientist evaluates organizational data, after which they recommend and provide specific actions that will help improve institutional performance, better customer engagement, and ultimately increase profits.
(3) Adaptation of best practices and focus on important issues: One of the responsibilities of a data scientist is to ensure that employees know and are familiar with the organization's analytics product. They prepare employees for success by demonstrating the effective use of the system to bring out understanding

and action. Once employees understand product capabilities, their focus can shift to addressing key business challenges.

(4) Identifying Opportunities: Data scientists can question existing processes and deductions to develop additional methods and analytical algorithms. The job requires them to be continuous and to improve the value found in organizational data.

(5) Making decisions based on measurable data-driven evidence: With the advent of data scientists, data collection and analysis from various channels has eliminated the need to take major risks. Data scientists create models using existing data that mimic a variety of possible actions. In this way, an organization can learn which method will bring the best business results.

(6) Evaluation of Decisions: Part of the battle involves making certain decisions and applying those changes. It is important to know how those decisions affected the organization, and this is where the data scientist comes in. It is beneficial to have someone to measure key metrics related to significant changes and measure their success.

(7) Identification and Filtering of the Target Audience: From Google Analytics to simple customer surveys, most companies will have at least one source of customer data collected. If it is not used properly, the data is useless. The importance of data science is based on the ability to take existing non-existent data alone and combine it with other data points to generate information that an organization can use to learn more about its customers and audiences. A data scientist can help by identifying key groups with accuracy and analyzing different data sources. With this in-depth knowledge, organizations can redeem services from customer groups and help profit margins grow.

(8) Hiring the Appropriate Talent for the Organization: Day-to-day learning is a daily activity in the employer's life, but that changes due to big data. With the amount of talent available through social media, corporate information, and job search websites, data science professionals can work on all of these data points to find candidates who best fit the needs of the organization.

Working Remotely

COVID-19 has forced companies to switch to remote operating systems as soon as possible. After all, remote implementation is broader and deeper than most organizations realize. In order to be successful in the long run, it takes a strategic approach and a significant investment to transform corporate culture.

Challenges of Remote Operation

The COVID-19 crisis is facing companies with a host of challenges in the short term. In order to keep the product, immediate acceptance of remote operation is required. However, this also brings some difficulties. The three major barriers to remote operating performance are the following:

(1) Remote operation is used in a structured way: Companies that see salvation in the rapid development of tools without structural repairs risk losing efficiency and frustration among staff and stakeholders. It is not enough to simply provide the necessary infrastructure and tools. Some companies already have the infrastructure and tools at the beginning of the crisis, but now they see their full potential (Jaffe, 2021).
(2) Ensuring long-term endurance: Where companies used to manage everything in the middle, now they have to plan everything from a distance. This requires great security changes and structural adjustment. This should not be taken lightly, as cyber-crime has increased during the outbreak. It also requires the commitment of each employee to navigate safely in his or her work tools (Jaffe, 2021).
(3) Balancing work and personal health of employees: With the immediate introduction of homework, employees experience a blurring of the line between work and private life. Early recipients point out that this has no effect on production but threatens cooperation and communication if left unchecked. Investing fully in the well-being of your employees is, therefore, an additional point of attention (Jaffe, 2021).

COVID-19 is forcing companies in bulk to use remote work control. Seeming to be very successful, but there are also many indications that the production and measurement of occupational health come under pressure. It will only be a real success if companies start to deal with long-distance construction. It is not advisable to take the wait-and-see approach because working remotely will be a new routine after a corona disaster.

4.2.2 Communication Technology

The impact of technology on work cannot be measured. Today it can be asinine to ignore how technology has changed business communication.

Ways Businesses Use Technology Communication Include

(1) For customers, using technology tools like bots is easier than ever to provide fast, efficient, and international customer service.
(2) Individually—using new platforms where employees can easily communicate. Using technology also helps to eradicate prejudice and discrimination within the workplace.
(3) In the industry—It is now much easier to communicate with people who work for other companies in their industry; people who sell products between companies or affiliates can easily communicate using technology tools.

The impact and use of technology on business communication are increasing and unpredictable. While it may seem small to change the type of message or presentation of partners, there are some real benefits that can affect all aspects of the business. It includes security, customer relationships and ultimately has a positive impact on the business line.

4.2.3 Digital Payments in Businesses

Businesses often view payment experience as just another touch of customer experience, but now technological change has proven that digital payments bring a ton of business opportunities to grow faster and more efficiently.

Consumer payments have seen the digital revolution become so intense that people can no longer set foot in the bank, and most transfers can be made through apps and online platforms. However, the same cannot be said of many departments of finance where most transactions are processed in a manual, paper-based manner.

Considering that a typical bank-to-bank payment cycle takes about 35 days to complete and almost half of the invoices are paid late, it is clear that unless businesses go beyond paper-based payments, they will be disrupted by efficiency.

It is a tool that is ready to accept secure, fast payment, which reduces costs to the company and increases company revenue. More customers: more and more customers prefer online shopping and make online payments with mobile wallets, credit/bank cards becoming more common for them.

4.3 Common People

4.3.1 Digital Applications

The WHO Academy mobile learning app is specifically designed for health professionals and is designed to enable them to increase their life-saving skills to combat COVID-19. It brings mobile access to the COVID-19 WHO resource resources, including up-to-date guidance, tools, training, and consultation meetings to support health workers in caring for patients infected with COVID-19 and protecting them as they carry out their critical work.

The app has a new feature that uses unpopular taxpayer augmented reality (AR) technology to help health workers cope with the critical challenge that many face every day during the epidemic—to remain virus-free while caring for patients. This 20-min study shows the correct techniques and sequences for installing and removing protective equipment, which is very important in keeping health workers safe.

The program is based on the needs of 22,000 international health workers in the WHO Academy's global survey conducted in March or April 2020, like prevention and control, case management, use of protective equipment and safety in the workplace, and risk communication and community involvement. It was a major shock to respondents who said a tangible reading of demand could help prepare for the challenges of COVID-19 and that they access information online through their phones and tablets (Hang et al., 2020).

4.3.2 Online Entertainment

Due to the growing reliance of entertainment companies on technology, there is now a clear growth in entertainment, especially on online platforms, such as streaming services, betting sites, multiplayer games and apps. Due to this, there is now a surge in growth when it comes to entertainment, especially on online platforms, such as streaming services, betting sites, multiplayer games, and mobile apps. To be precise, the entertainment and media industry is expected to generate $ 2.2 billion by 2021 through online entertainment, such as broadcast services, which has led to this significant growth. This is due to the growing popularity of online entertainment–180 million Americans currently have access to video content. While you may not believe that the growth of online entertainment reduces the need for physical and traditional venues, this is not the case, many of these companies are still finding growth as well (Affinito et al., 2020).

4.3.3 Digital Payments

Common People have stopped accepting payment for the paperwork with too many complaints:

(1) Paper checks shorten the payment cycle and often create delays.
(2) Accepting checks puts businesses at greater risk of fraud.
(3) Reconciliation is very costly and flawed.
(4) There is no guarantee that the check will be linked to the account with sufficient funds.
(5) Customers can place stop payments or close an account.
(6) Spending time driving to the bank because remote photography is not provided.
(7) It is difficult to follow an incoming check or know when the payment will arrive.

With digital payments, you can process payments very quickly, even on the same day, allowing faster deposits and automatic reconciliations, which are proven to eliminate data entry errors and ensure your books are accurate.

4.3.4 Communication Technology

Benefits of Technology in Business Communication.

(1) Talking in real-time: It does not mean that it is much easier to communicate quickly using apps than many older methods, such as telephone and fax. Using

the benefits of technology at work, you may be able to talk to people in real-time, whether it is more than just a messaging app or a video call on your PC, tablet, or interactive display. There are a variety of applications that are suitable for internal communication and can be used to communicate with people in the company from elsewhere.

(2) Remote communication: One of the great advantages of communication technology in the workplace is how easy it is now to communicate with remote employees. The future work environment is highly dependent on remote employees, so it is important to have forums to contact. This would not have happened if businesses had relied on home telephony and fax machines. These days, messaging apps and a huge amount of mobile data make it very possible.

(3) Increase in Security: The more the usage of technology is, the more secure it becomes. Using emails, especially when they were sent across the office, without adequate security was a disaster risk. These days, most apps use end-to-end encryption, meaning that the sender and recipient can only read messages. This means that emails are less likely to be hacked. In addition, external systems often require a password to log in and have features such as encryption, making them more secure.

(4) Improves customer relationships: Companies that have used technology to communicate at work are jumping forward in terms of customer relationships. Many businesses have used the use of bots that can quickly and easily answer customer questions. Video messaging apps like Google Hangouts are great tools that can help businesses communicate with customers around the world. Good planning is another great way to use technology to build a good foundation for customer relationships and customer relationships. People who frequently use business services will download the app and use it to communicate with you through comments, questions, or improvements.

(5) Versatile in nature: As well as sending messages and to-do lists, business communication methods include sharing ideas. This can be done in a variety of ways, which are very popular. They are a great way to share business ideas using different types of communication technology. They usually include business development programs or solutions to any problems. The presentations have come a long way since PowerPoint. These days, they can be introduced in a variety of ways. Another great feature is the interactive display, a different kind of interactive whiteboard in the private room. This creates an intimate setting where viewers can easily see the screen and know how much detail there is. Collaborative consultation is sensitive and can be dragged into it, making them a great way to communicate different messages.

4.3.5 Television

Accurate and timely prediction of program popularity is critical for content providers, advertisers, and TV broadcasters. This information can benefit operators in TV program purchases and can help advertisers develop effective marketing strategies.

In addition, in case of any technical issue, a more direct way of predicting the popularity of a program can improve the entire streaming system, such as a content delivery strategy and Cache strategy. Several types of predictions have been suggested based on YouTube and Twitter. However, existing prediction methods often require a large number of samples and a long training period, and the predictions' accuracy is incorrect in systems that meet high peaks or sharp declines in popularity.

4.3.6 Online Shopping

As the world responds to the coronavirus epidemic (COVID-19), we are seeing a dramatic shift from physical to online shopping. Consumers are relying on the digital world more than ever and businesses are forced to change their strategies and move to digital transformation with greater urgency than ever before.

There has been a 6% increase in sales of cosmetics and personal care products, 1% increase in sales of digital entertainment, less than one percent increase in sales of agro-food and beverage products, 2% increase in sales of fashion and accessory products, 9% increase in sales of pharmaceutical and health products, 4% increase in sales of media and books (both digital and online) products, 10% increase in sales of ICT/electronic goods, 7% increase in sales of home furniture and household products, 8% increase in sales of education and online courses, a9% increase in tools on online platforms. On the other hand, there has been an 11% decrease in sales of agro-food and beverage products, 23% decrease in sales of pharmaceutical and health products, 27% decrease in sales of media and books (both digital and online) products, 28% decrease in sales of digital entertainment, 29% decrease in sales of education and online courses, 32% decrease in sales of cosmetics and personal care products, 35% decrease in sales of tools, 48% decrease in sales of ICT/electronic goods, 53% decrease in sales of home furniture and household products, 75% decrease in sales of Tourism on offline entrepreneurship (Prasetyo & Fuente, 2020).

Nowadays people's lifestyle is different. People feel uncomfortable spending a lot of time going to crowded markets. Therefore, E-Shopping is a blessing as it saves a lot of time. Online shopping is a process in which consumers purchase direct goods, services etc., from the seller without an online mediation service. Buyers can visit web stores from their homes and shop as a computer sit-in.

Online websites are usually available 24 h a day and most consumers have internet access both at work and at home. So it is very easy for them to shop online. One of the great things about online shopping, especially during the holidays, is reducing the need to wait in long lines or search the store for something. A variety of goods are available online (Galhotra & Dewan, 2020).

4.3.7 Work from Home

The COVID-19 epidemic and its after-effects have made work from home a necessity in many industries. Working from home or away from home is a relatively new phenomenon and looks set to continue working for years to come because of the epidemic. Working from home is nothing new as there are many self-employed people who work for many organizations far away. But now, it has been accepted by organizations for almost all of its employees.

It is a situation that has been playing out across the country as the number of coronavirus cases increases and many employers are committing themselves to social media practices that health officials say will help slow the spread of COVID-19, a respiratory illness caused by a virus. High-tech employers such as Apple, Google, Facebook and Microsoft—which have a large population based in Seattle and other cities which are most affected by the epidemic—were among the first to strengthen the remote applications of many or all of their employees worldwide.

67% of employers are taking steps to allow employees to work from home who do not normally do so. 36% of the employees were actively encouraging other employees to work from home. 42% of the employees were encouraging other employees to work from home on a case-by-case basis. Only 6% of the employees were encouraging other employees to work from home in hot spots (Shareena & Shahid, 2020).

Work from home comes with its negatives, too, such as:

(1) Difficulties in the separation of health and homework: Working from home is not a dream come true. In fact, it could even harm your mind. Having time to go home from work allows you to avoid the pressures that your work may cause you mentally. When you work from home, that transition is no more, and the line between work life and home life is blurred.
(2) Decreased Public Health: While working from home may not be a distraction from other people in the office, it can be difficult in your social life. When you spend eight hours of your day indoors without interacting with people, you will no doubt begin to feel lonely.
(3) Difficulty with Motivation: You may say goodbye with a motive when you decide to work in the comfort of your home. When you do not have to get up early to get to the office, you can bet that you will choose to sleep. When you don't need to get dressed, you can make sure you stay up all night wearing pajamas. When you do not deal with normal social norms in the workplace, you will find that you are less likely to do things that fit your motivation levels.
(4) Lack of cooperation: New evidence suggests that working from home can negatively affect your old thinking ability. When people with different mental processes surround you, it is easy to be creative and come up with great ideas! By working collaboratively and hitting ideas on your employees, you will no doubt be very successful in finding creative solutions (Shahid & Mahammad, 2020).

22% of the employees found it difficult to unplug after work. 19% of the employees felt loneliness. 17% of the employees found it hard to collaborate and communicate with other employees. 10% of the employees found it hard to work because of the distractions at home. 8% of the employees found it difficult to work with other employees due to them being in different time zones. 8% of the employees found it hard to stay motivated. 7% of the employees found it hard to take vacation time. 3%

of the employees found it hard to find a reliable WIFI connection. In contrast, the other 4% of the employees have other issues (ShareenaShahid, 2020).

4.4 Police Force

4.4.1 Drones (Robotics)

Police drones are equipped with high-definition cameras with powerful zoom capabilities and thermal cameras, enabling them to see during the night like in daylight. These drones can take off almost anywhere or keep hovering for a long time. From the air, flying high and away from its target, the police drone provides an eagle vision to the officers. Ready for immediate take-off at any time, the Police drone can be quickly on-site without involving any greater resources. This way, the manned Police helicopters remain available for other missions for which the transport of people by air is essential. By observing from above, the drone can limit the risk of exposure of Police officers and thus save human lives. Police operation headquarters are in permanent contact with the drone pilots so they can guide the ground special units and squads. The Police drone pilots team is available 24 h a day, 7 days a week. The high mobility of the drone allows the Police officers to get to the scene very rapidly and have the aircraft airborne almost immediately.

In the United States of America itself, there has been a significant rise in the number of drones between 2016 to 2021, from 110 to 996, which cost 0.6 billion dollars in 2016 to 3.6 billion dollars in 2021. There has been a prediction which states a further sharp rise in the number of drones in the future, between 2021 to 2025, from 1996 to 2679, which cost 3.6 billion dollars in 2021 to 12.6 billion dollars in 2025 (Chamola et al., 2020).

In many countries, police drones are being used to monitor the public's compliance with lockdown regulations and encourage or enforce social distancing. These are central strategies in the fight against the virus involving gathering data about people, such as the locations they visit. The use of drone surveillance must be carefully balanced against protecting our human rights. It should undergo robust and independent ethics reviews and be subjected to human rights impact assessment. Without this, drones could become a key building block of a surveillance architecture that may last well into the future (Chamola et al., 2020).

4.4.2 Communication Technology

Communication technology is important in many industries and one of the most important areas in the field of public safety. For those who are committed to security and performance, having the right equipment is a daily necessity. And when it comes to police communication technology, in particular, there are a number of important communication tools that can give officers trust, real-time information that is always informative and secure and responsive.

Two-Way Radios

The two radio stations keep officials communicating in the most challenging areas in the camp and at work. Two-way radios are designed to be rigid and reliable and work where cell phones are not available—even in times of emergency, power outages, and overcrowded cell towers. The two radios also use push-to-talk technology, allowing faster communication with individuals and multiple groups simultaneously. Two-way radios are an excellent example of police communication technology because they are lightweight and powerful. And they are compact, easy to carry with other gears.

Page Input Programs

Another police technology is paging. While many pagers have been replaced by SMS messages, today's page solutions offer advantages over mobile phones. Basically, they send messages using high-quality radio signals, which means they can reach far, require fewer senders, and have less interference than cell phones. Sources say that pager messages can often go to remote areas where cell phones and SMS text messages can. Additionally, pager messages can pass quickly - in just a few seconds, as opposed to the minutes, hours, or days. SMS messages can take when networks are crowded.

Police Body Cameras

Wearing body cameras is another police communication technology that could improve performance and is becoming commonplace in police departments across the country. While there are some privacy issues and potential technical problems, these cameras offer law enforcement agencies many benefits, such as reducing complaints about official conduct, producing evidence, preventing false accusations, and increasing transparency. And they can increase the similarities between the police and the citizens.

Emergency and 911 Communications

Ultimately, tool officers who find importance in their work are the 911 call solution that connects the community to emergency services. The 911 quality system allows emergency response communications to be centralized and cleaned, so dispatches can control existing resources and use them efficiently. The app can also be integrated between multiple social security agencies and can include voice, video and data, including a map and GPS. This means that police can receive and transfer real-time information from the scene and obtain important information when answering calls.

Many public safety organizations use 911 communication technology known as Emergency Call Works. This application uses a single application to locate 911 calls, manage call locations on a map, and send field resources. It can be useful to the police because of software-focused means it can be accessed from anywhere. This can add to the potential savings of public safety structures as they do not need to invest in additional infrastructure, and problems can be identified and remedied.

Police communication technology is an important ingredient in helping law enforcement protect the public and keep them safe. Also, the right investment can keep your police officers strong.

4.4.3 Digital Applications

Many applications have been made or submitted, with official government funding in some areas and elsewhere. Several frameworks for creating contact tracking applications have been developed, and privacy concerns have been raised, especially regarding applications based on location tracking around the app's users.

Application-based contact tracking is partly interesting because the spread of coronavirus is intense. People with the virus can transmit the virus a few days before they develop symptoms, and it can take a few days for community health investigators to learn about the case and confirm its diagnosis. These groups then had very little time to follow the traditional trail: interviewing an infected person, tracking all the contacts they recently remembered and making those people isolated before they also transmitted the virus.

Many countries, including India, South Africa, China, Australia, Russia and Canada, just to name a few, have been using contact tracing. Other countries like the United States of America and Alaska have their jurisdictions under development as of March 2020 (Alhasan & Hasaneen, 2021).

Some countries have used network-based location tracking instead of apps, eliminating the need to download the app and the ability to avoid tracking. In Israel, network-based tracking is allowed. Network-based solutions that have access to immature local data have significant potential privacy issues. However, not all programs with centralized servers require access to personal location information; it has created many privacy programs that use only central servers to communicate.

A basic test is whether the sensors of the phone can select the types of contacts that can transmit the virus. The app must measure the distance between two people depending on the strength of the Bluetooth backup from one phone to another. Investigators are now conducting experiments to find out how walls and other obstacles reduce signal strength. Thus the application calculates the risk (Alhasan & Hasaneen, 2021).

Educational Sector

Students, their parents, and teachers around the world are experiencing the extraordinary effect of the novel coronavirus as schools are closed and segregation measures are ordered to deal with a global epidemic. While governments and health officials are doing everything they can to reduce the spread of the disease, global education programs are working together to provide a completely comprisable and comprehensive education system for all at this critical time. The aim is to assist everyone in providing quality education to all and at these different times.

Online Entertainment

Students, their parents, and teachers worldwide are facing the novel coronavirus's shocking effect as schools are closed and discriminatory measures are being ordered to deal with a global epidemic. While governments and health authorities are doing everything they can to reduce the spread of the disease, global education programs are working together to provide a comprehensive and comprehensive education system for all at this critical time. Our job is to help everyone provide quality education for all and at these different times.

There has been a 27% increase in the watch time of Facebook, 16% increase in the watch time of Netflix and 15.3% increase in the watch time of YouTube on websites in the United States of America. On the other hand, there has been only 1.1% increase in the watch time of Facebook, only 0.3% increase in the watch time of Netflix and 4.5% decrease in the watch time of YouTube on apps in the United States of America. There has been a 12.4% increase in the usage of Google Duo app, 73.3% increase in the usage of the Nextdoor.com website, 79.4% increase in the usage of the Houseparty app since the pandemic emerged. There has been 19.7% increase in the usage of Twitch.tv website, 15.4% increase in the usage of the TikTok app, and 40.2% decrease in the usage of the ESPN.com website since the pandemic emerged (Alexopoulos et al., 2020).

The epidemic has led to layoffs in some areas of media and entertainment such as print and television and mainly sports issues as advertisers cut back on spending. As monetization, especially capital expenditure, comes under pressure, financial management and profit protection with greater integration of technology is likely to find strategic value in media and entertainment companies. While long-term outcomes are yet to come, subsequent media and entertainment themes may be focused on as the reality of post-coronavirus becomes clearer.

4.4.4 Digital Applications

Education is always in jeopardy because it is vital to economic prosperity and human resource development. Mobile learning has also overcome the challenges of education. Collaborative and customized learning is a resource that students find in mobile apps. For students, educational-based apps provide ease of use by helping them achieve more in less time. It can be said that mobile learning apps will prove to be the best friends of students in times of need.

(1) Artificial Intelligence: From watching TV, dating, or shopping, Artificial Intelligence (AI) has become a part of our lives. AI has a huge impact on both the educational and developmental benefits of mobile applications. This over-the-top technology will reduce classroom instructions but will improve it in many ways. Let's go through a few benefits of using AI in the field of education:
(2) Customize student reading: AI technology encompasses one of the biggest benefits (i.e. personalization) in the education industry. For teachers, it is difficult to pay attention to every student in the class. However, Artificial Intelligence can help to adapt to each student's reading.
(3) Global cooperation is possible: With the use of AI, schools will be able to create a global classroom. Artificial Intelligence technology will take the education industry higher. For example, there will be no need for students to be physically present, so absenteeism due to illness or other reasons is not a problem. This is because AI will connect students to the visual classroom. In addition to institutions such as virtual classes, AI also provides peer-to-peer communication for students who are miles away.
(4) Conversation dialogues: Chatbots is another high-tech mobile technology platform that will empower online education. Chatbots not only provide the learning environment itself but also transform conversations into conversational conversations. It will also convert conversations into a series of messages. As a result of discussions to make learning easier for students. This game-changing technology will test students' level of comprehension. In short, chatbots provide learning information designed for students.
(5) Virtual Reality: The next trend in hyped mobile technology is Virtual Reality. It has come and given great impetus to internet technology. Thanks to VR, it took a learning process beyond the learning environment. Speaking of medical students, VR has already made things easier; allows live surgical streaming views. VR offers something unusual by giving them what is happening in the real world, which they would not find until late at night in their training.
(6) UI: The combination of reading and fun can drive more engagement with students. Strengthening has the potential to turn to learn into fun, thus leading to better student engagement. To improve the level of student harassment, it is important to increase productivity. Therefore, we can say that the explosion of gamification will give a new edge to online learning.

There has been a sharp rise in the usage of Zoom, Google Classroom, Microsoft Teams, Hangouts, etcetera since the pandemic emerged as working professionals and common people began using them as their main mode for communication (Alexopoulos et al., 2020).

Powerful Solutions Given Student Learning Programs.

(1) Provides Formal and Intelligent Reading: Educational apps offer systematic and intelligent learning. Educational programs are organized in such a way that students can flow.
(2) Improved communication: Educational applications can make students more active and create better engagement between parents and children. The tendency to interact is also increasing in children with mobile learning apps.
(3) Bridge Gap between Parent and Teacher: Normal parent-teacher interactions have never happened for a number of reasons, such as strict schedules for teachers and parents. It is not always possible to maintain contact through physical contact. Anyway, we have educational apps now. Teachers can now pay attention to parental questions anytime and anywhere with the educational app. Educational programs promote visibility in the development of children in their schools.
(4) Child Development: There is nothing more important than tracking your child's progress. Along with development, mobile learning apps let you visualize how each app helps your child develop their skills, such as learning new languages, math, and much more.
(5) E-Books and Online Learning Materials: Mobile educational apps come with many benefits for students, as it saves a lot of money to buy reading materials in stores and libraries. Mobile learning apps allow students to learn lessons from an educational app with just a few simple clicks. Students can also get mixed-up learning through educational apps.
(6) 24/7 Availability: The great thing about mobile apps is that they are available to students and they can learn at the right time.
(7) Encourages self-study: In the traditional education system, self-study is unprecedented. Educational apps come with a lot of resources so that students can interact with them happily.

4.4.5 Digital Payments

When it comes to finances and payments, the educational environment is set to address some of the key challenges in the coming years. Each, in its own way, will present an obstacle to success that needs to be overcome:

(1) High fees: The pressure of such a large amount of money can be seen. Loans have always been avenue students who have gone down to distribute the cost of their diploma, but this is gradually declining due to the high-interest rates involved. Therefore, it is becoming increasingly important for universities and other education providers to reduce the financial impact of funding their studies, either by providing the opportunity to pay in installments or finding cheaper ways to provide education. This is an area where eLearning can help reduce tuition costs by eliminating the need for a teacher or class.
(2) International Student Surge: While providing an exciting new source of income for universities, catering to this international market has its challenges. It is common for student families to pay for their children's education and books and

other materials from their home country. This means that education providers need to ensure that they are able to empower students to pay for their studies from abroad.

(3) Location Divided by Payments Methods: Regardless of where they come from, students want to be able to pay using the method they are familiar with—in this fast-paced global payment system, these methods can vary from region to region. Contrary to what has happened in the past, education providers should give students and their families the opportunity to choose payment methods to pay for their studies financially using the technology they are familiar with. Also, this offers a wide range of educational platforms that can provide an opportunity for teachers to offer a simple payment system anywhere in the world, providing a choice of payment methods to meet local needs.

Therefore, it is profitable to pay for education online because it is not only safe but also reliable and not an inconsistent way of doing transactions.

4.4.6 A Study from Home or Distance Learning

The majority of university students still prefer to study at their own home. This allows the students to create a particular learning space that is possible to adapt to every career, study, or need.

Advantages of Studying at Home

1. The biggest advantage of studying at home is that the students have everything at their disposal without leaving the chair. In addition to books and notes, students can use the computer, the printer, the scanner, etc.
2. The comfort of the room also allows the students to eat and drink while they study. Although it is always not recommended to take any food that stains, students can accompany the day with nuts, chopped fruit, juices, milkshakes, etc.
3. The solitude of the room allows students to read aloud and repeat the syllabuses, being able to perform simulations and practices of the presentations.
4. It does not need transferences or loses a minute to go to eat, since students can stop studying and immediately take the plate.
5. Students can take a nap of about 15 to 20 min to replenish their strength and regain the energy they need to continue studying.
6. Students can study how they want, from the comfort of home. In addition, not depending on a schedule, students can take advantage of the nights or mornings to study.

Disadvantages of Studying at Home

1. Students will need to have a number of elements to start the study, such as a large table and an ergonomic chair. If the student does not have space in his or her room, he or she may need to move his or her study area to the living room.
2. Students will not have anyone around them to ask a question or make an appointment. In this case, the students will have to pick up the phone to call a colleague, send an email to the teacher, or wait for the next day to discuss it with a partner.
3. Whiling time is often common among university students who study at home, since individual learning requires some constancy, discipline, and responsibility.
4. Studying all day at home can be quite anti-productive and students may need some time to enjoy the outdoors. In addition, being in the room for prolonged periods can increase the temptation of lying in bed, surfing social networks, or taking a nap between the topics.
5. Silence is essential to study while concentrating properly and a home is not the oasis of tranquillity that a student needs. If students cannot focus on the subject, perhaps they may also entertain more because of the easy access to the Internet, books, video games, etc.

5 Proposed Method and Implementation

The usage of CNN modeling to classify if a chest x-ray image is healthy or not. The function performs multiclass classification such as normal, virus, and bacteria. To solve this problem, CNN modeling (Convolutional Neural Network) is used due to its excellent image-splitting ability.

5.1 System Flow

The system flow, which is the working of the implemented program, is shown in Fig. 3.

The input is accepted as chest x-ray images, after which the images and their labels are renamed. The images are then pre-processed by resizing to 200 by 200 pixels and grey scaling. The labels are then merged with their respective images, which are then reshaped and augmented. Then the input shape of the neural network is then defined, and a two-dimensional convolution layer is generated. Each input layer is connected to several convolution pooling layer pairs before flattening and getting connected to the dense layers. The CNN model is now ready to be created and the last layer will have three neurons. Then the model is compiled using a categorical cross-entropy optimizer and an adam optimizer, after which the model is trained and fitted (Priya et al., 2020). The summary is now generated and test data is used for

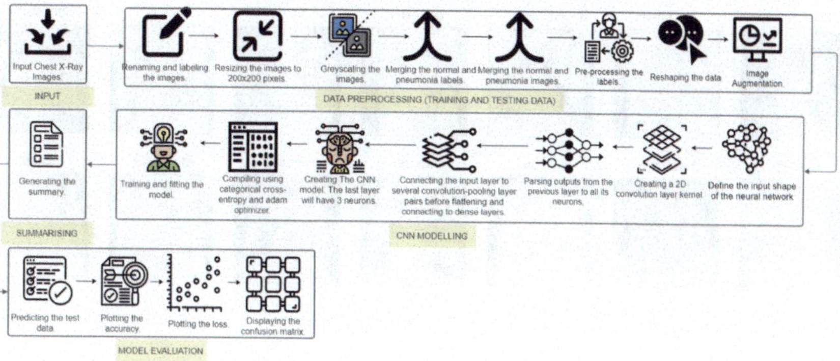

Fig. 3 System flow of the proposed idea

the prediction. After the prediction is made for all the test data, the output prediction is compared to the actual labels and the accuracy and loss plots are generated and finally, the confusion matrix is generated (Abirami & Palanikumar, 2021).

5.1.1 Dataset Used

The dataset used in this project (1.3 GB) could be found and installed from the Kaggle link mentioned below, https://www.kaggle.com/paultimothymooney/chest-xray-pneumonia.

The constituents and description of these subfolders are mentioned below:

This is the main folder which contains three main subfolders named test, train and val (validation).

- The training dataset is generally the largest dataset as the model needs to be created from the data it provides.
- The testing dataset is generally smaller than the training dataset but larger than the validation dataset as this generally provides the accuracy of the model by testing the model.
- The validation dataset generally is the smallest dataset and provides the validation to the model.

The training dataset comprises Normal or healthy lungs (1341 images) with the existence of a total of 766 people's x-rays and pneumonia or people having unhealthy lungs (3875 images), among which 1345 and 2530 samples for virus and bacteria, respectively. There exists singular or multiple x-rays images of individual people. In the second case, there are in total 1945 people's x-rays. In comparison, the test dataset comprises two output classes such as healthy lungs (234 images) and people having unhealthy lungs (390 images) (Dhiman & Chang, 2021).

The program reads the data and then, depending on the images' characteristics, the model is made, which can now classify if a pair of lungs are healthy or unhealthy

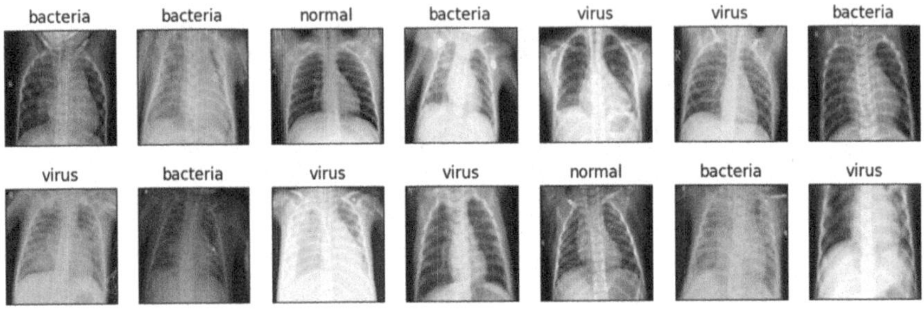

Fig. 4 Sample Input Data

and if they are unhealthy, then does the person have a virus or bacterial infection. The contents of the validation dataset are used to validate the classifier after it has been made, and it could be considered as a part of the training dataset (Abdel-Basset et al., 2020, 2021). This subfolder contains two more sub-sub folders, such as Normal (8 images) and Pneumonia (8 images). All the images in this sub-sub folder are x-rays of people having healthy and unhealthy lungs (bacteria or virus). It prevents overfitting of the classifier and helps to tune the parameters of the model (Yonghang et al., 2021).

5.1.2 Implementation

These are input and the results in the form of plots of the CNN model created from the dataset mentioned above.

5.1.3 Input Data (Images)

The input data, chest X-rays having a bacterial infection or a virus, or normal lung images are shown in Fig. 4.

5.1.4 Training and Building the CNN Model

Figure 5 shows that the loss value keeps increasing and becomes constant, marking the end of training and building the CNN model (Fig. 6).

5.1.5 Result and Analysis

Accuracy, Misclassification rate, Precision, Sensitivity (Recall), Specificity and F1-score can be calculated using the formulae mentioned below:

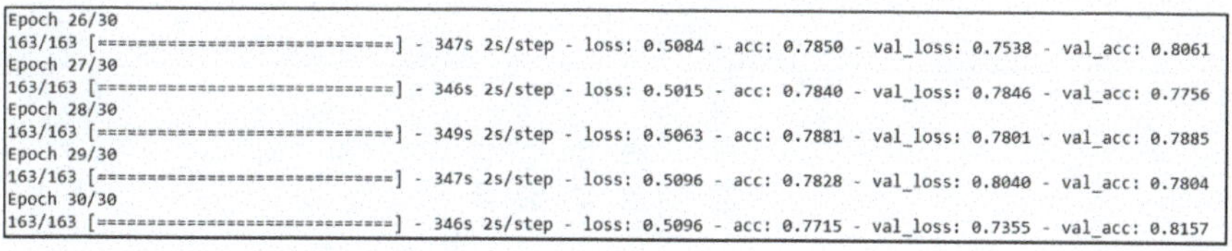

```
Epoch 26/30
163/163 [==============================] - 347s 2s/step - loss: 0.5084 - acc: 0.7850 - val_loss: 0.7538 - val_acc: 0.8061
Epoch 27/30
163/163 [==============================] - 346s 2s/step - loss: 0.5015 - acc: 0.7840 - val_loss: 0.7846 - val_acc: 0.7756
Epoch 28/30
163/163 [==============================] - 349s 2s/step - loss: 0.5063 - acc: 0.7881 - val_loss: 0.7801 - val_acc: 0.7885
Epoch 29/30
163/163 [==============================] - 347s 2s/step - loss: 0.5096 - acc: 0.7828 - val_loss: 0.8040 - val_acc: 0.7804
Epoch 30/30
163/163 [==============================] - 346s 2s/step - loss: 0.5096 - acc: 0.7715 - val_loss: 0.7355 - val_acc: 0.8157
```

Fig. 5 Training results of CNN model

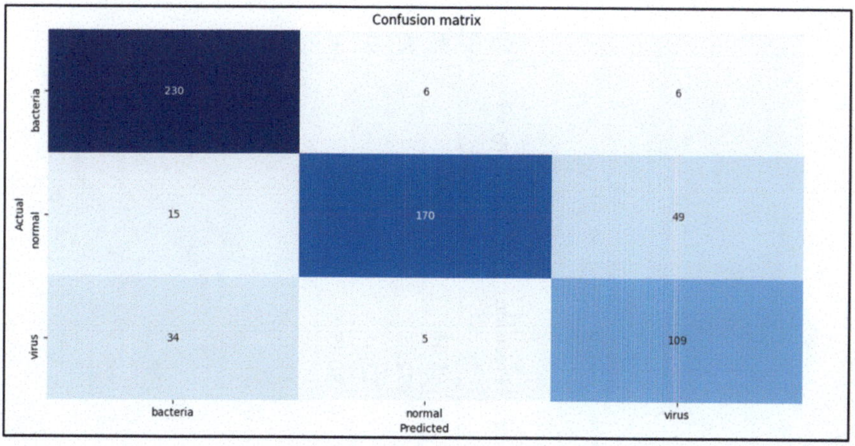

Fig. 6 Three variable (normal, bacteria and virus) confusion matrix

- Accuracy: (TP + TN) / (TP + TN + FP + FN):
- Misclassification rate: (FP + FN) / (TP + TN + FP + FN) or (1 - Accuracy):
- Precision: (TP) / (TP + FP):
- Sensitivity: (TP) / (TP + FN):
- Specificity: (TN) / (TN + FP):
- F1-score: (2 × Precision x Sensitivity) / (Precision + Sensitivity) or (2 × TP) / ((2 × TP) + FP + FN):

Table 2 shows the count of the true positive, true negative, false positive and false negative, the accuracy, the misclassification rate, the precision, the sensitivity, the specificity and the F1-score.

5.1.6 Accuracy

Figure 7 shows the accuracy, which increases with respect to the amount of data used to create the model. It compares the accuracy with and without validation and it is clearly visible how important validation is. Validation increases the accuracy from around 78 to 87.71%.

Table 2 Evaluation measures

Variable	TP	TN	FP	FN	Accuracy (%)	Misclassification rate (%)	Precision (%)	Sensitivity (%)	Specificity (%)	F1-score (%)
Bacteria:	230	333	12	49	90.22	9.78	95.04	82.44	96.52	88.29
Normal:	170	379	64	11	87.98	12.02	72.65	93.92	85.55	81.93
Virus:	109	421	39	55	84.94	15.06	73.65	66.46	91.52	69.87
Average (Mean)					87.71	12.29	80.45	80.94	91.20	80.03

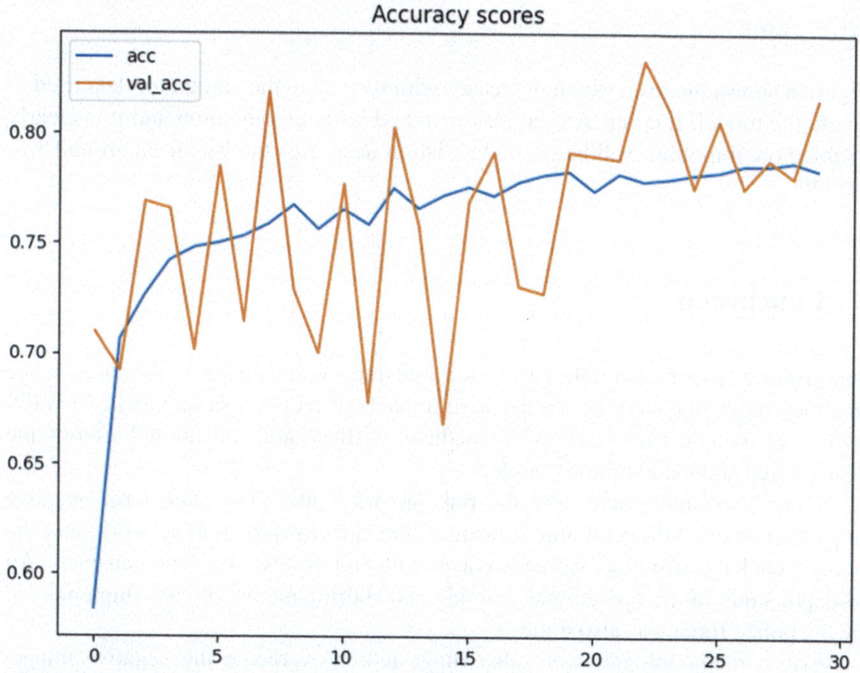

Fig. 7 Accuracy plot (with and without validation)

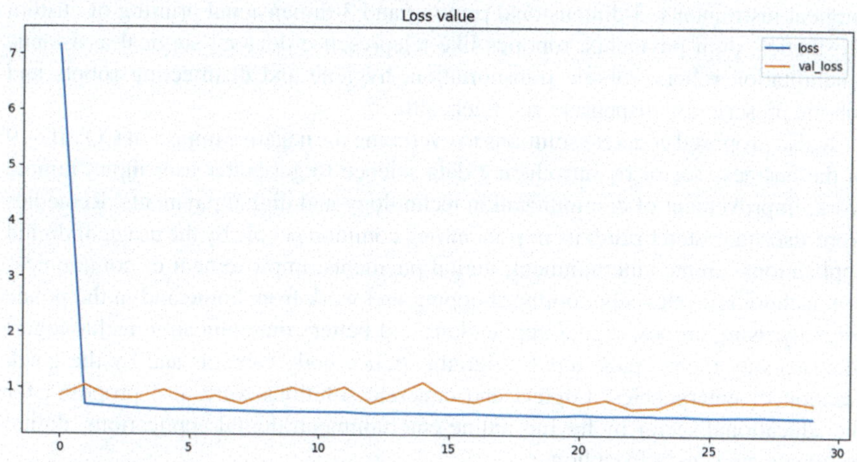

Fig. 8 Loss plot (with and without validation)

5.1.7 Loss

Figure 8 shows the loss, which decreases with respect to the amount of data used to create the model. It compares the loss with and without validation and it is clearly visible how important validation is. Validation decreases the loss from around 1 to 0.5096.

6 Conclusion

The proposed paper studies the risk factors and impacts and discusses the data science methodologies to overcome the negative impacts or reduce risk factors of COVID-19 in various industries, such as the medical, business and educational sectors, the police force and the common people.

It was also highlighted why the risk factors matter. The short-term negative impacts, revenue fall, economic downturn, the shift towards remote work, and the bounce-back of businesses were also dealt with and discussed comprehensively. An in-depth study of the operational, training and staffing methodologies implemented by the police force was also done.

Various methodologies were also suggested to overcome the negative impact of COVID-19 in the medical sector by implementing medical training techniques, genome sequencing, 3-dimensional printing like tissue and organ bioprinting, surgical instruments, 3-dimensional printing and 3-dimensional printing of custom prosthetics, digit payments, robotics like telepresence doctors, surgical assistants, rehabilitation robots, robotic transportation, hygiene and disinfecting robots and robotic prescription dispensers and telehealth.

It also proposed concrete solutions to overcome the negative impact of COVID-19 in the business sector by introducing data science to get better user input, remote work, improvement of communication technology and digital payments. Remedies were also suggested to curb its impact among common people by the usage of digital applications, online entertainment, digital payments, improvement of communication technology, television, online shopping and work from home and in the police force by using drones, digital applications and better communication technologies like two-way radios, page input programs, police body cameras and by the quick reaction of emergencies. Logical and concrete solutions were also proposed for the educational sector by having online entertainment, digital applications, digital payments, and study from home.

The CNN model was created and demonstrated how it classifies if a pair of lungs are normal or infected with pneumonia. The CNN model gave 87.71% accuracy, 12.29% miscalculation rate, 80.45% precision, 80.94% sensitivity, 91.20% specificity, 80.03percent F-1 value against the testing dataset having a loss value of 0.5096.

References

Abdel-Basset, M., Chang, V., & Mohamed, R. (2020). HSMA_WOA: A hybrid novel Slime mould algorithm with whale optimization algorithm for tackling the image segmentation problem of chest X-ray images. *Applied Soft Computing, 95*, 106642. https://doi.org/10.1016/j.asoc.2020.106642

Abdel-Basset, M., Chang, V., Hawash, H., Chakrabortty, R. K., & Ryan, M. (2021). FSS-2019-nCov: A deep learning architecture for semi-supervised few-shot segmentation of COVID-19 infection. *Knowledge-Based Systems, 212*. 106647, ISSN 0950-7051, https://doi.org/10.1016/j.knosys.2020.106647.

Abirami A., Palanikumar S. (2021) Proactive network packet classification using artificial intelligence. In: Misra S., Kumar Tyagi A. (eds) *Artificial Intelligence for Cyber Security: Methods, Issues and Possible Horizons or Opportunities. Studies in Computational Intelligence*,972. Springer, Cham. https://doi.org/10.1007/978-3-030-72236-4_7

Affinito, A., Botta, A., Ventre G. (2020). The impact of covid on network utilization: In analysis on domain popularity. In: *IEEE 25th International Workshop on Computer Aided Modeling and Design of Communication Links and Networks (CAMAD)* (pp. 1–6) https://doi.org/10.1109/CAMAD50429.2020.9209302.

Alexopoulos, Hudson, Arion., Otenigbagbe, Jake., Oluwatomisin. (2020). The use of digital applications and COVID-19. *Community Mental Health Journal*. 56. https://doi.org/10.1007/s10597-020-00689-2.

Alhasan, M., & Hasaneen, M. (2021). Digital Imaging, Technologies and Artificial Intelligence Applications during COVID-19 pandemic. *Computerized Medical Imaging and Graphics., 91*, 101933. https://doi.org/10.1016/j.compmedimag.2021.101933

Chamola, V., Hassija, V., Gupta, V., Guizani, M., (2020). A comprehensive review of the COVID-19 pandemic and the role of IoT, drones, AI, blockchain, and 5G in managing its impact. In *IEEE Access* **8** (90225–90265) https://doi.org/10.1109/ACCESS.2020.2992341.

Gaurav Dhiman, Victor Chang, Krishna Kant Singh Achyut Shankar (2021) .ADOPT: automatic deep learning and optimization-based approach for detection of novel coronavirus COVID-19 disease using X-ray images, *Journal of Biomolecular Structure and Dynamics*, DOI: https://doi.org/10.1080/07391102.2021.1875049

Galhotra, B., Dewan, A. (2020). Impact of COVID-19 on digital platforms and change in E-commerce shopping trends. In *Fourth International Conference on I-SMAC (IoT in Social, Mobile, Analytics and Cloud) (I-SMAC)* (pp. 861–866) https://doi.org/10.1109/I-SMAC49090.2020.9243379.

Gopisetty, S. (2020). Global pandemic: Business model impact on enterprises reTHINK, reIMAGINE, reINVENT businesses. In *2020 IEEE Second International Conference on Cognitive Machine Intelligence (CogMI)* (pp. 114–120). https://doi.org/10.1109/CogMI50398.2020.00024.

Hang, A., Dascalu, M. I., Stanica, I. (2020). Contact tracing app for containing diseases spread. In *Zooming Innovation in Consumer Technologies Conference (ZINC)* (pp. 216–217). https://doi.org/10.1109/ZINC50678.2020.9161820.

Hořínka, J., & Kontriková, L. (2020). Covid-19 pandemic impact on the aviation future. In *2020 New trends in aviation development (NTAD)* (pp. 99–104). https://doi.org/10.1109/NTAD51447.2020.9379075.

Jaffe, S. (2021). Work from home during and after COVID-19. In *IEEE/ACM 8th International Workshop on Software Engineering Research and Industrial Practice (SER&IP)* (pp. 28–28). https://doi.org/10.1109/SER-IP52554.2021.00012.

Khanna, D., Prasad, A.(2020). Problems faced by students and teachers during online education due to COVID-19 and how to resolve them. In *6th International Conference on Education and Technology (ICET)* (pp. 32–35). https://doi.org/10.1109/ICET51153.2020.9276625.

Nabi, S., & Mishra, V. P. (2021). Analysis and impact of COVID-19 on economy and organization. In *2021 International Conference on Computational Intelligence and Knowledge Economy (ICCIKE)* (pp. 219–224). https://doi.org/10.1109/ICCIKE51210.2021.9410734.

Pott, P., Weiser, P., Scharf, H., & Schwarz, M. (2004). A gearing mechanism with 4 degrees of freedom for robotic applications in medicine. *BiomedizinischeTechnik. Biomedical Engineering., 49*, 177–180. https://doi.org/10.1515/BMT.2004.033

Prasetyo, Y. T., Fuente, D. G. D. D. (2020). Determinant factors affecting customer satisfaction among filipinos in lazada online shopping during COVID-19 pandemic: A structural equation modeling approach. In *7th International Conference on Frontiers of Industrial Engineering (ICFIE)* (pp. 48–52) https://doi.org/10.1109/ICFIE50845.2020.9266734.

Priya, R. L., Jinny, S. V. (2021). Elderly healthcare system for chronic ailments using machine learning techniques—a review. *Iraqi Journal of Science* 62(9), (3138–3151). https://doi.org/10.24996/ijs.2021.62.9.29.

Priya, R L., Jinny, S., Vinila , Mate.(2020) Y.V. Early prediction model for coronary heart disease using genetic algorithms, hyper-parameter optimization and machine learning techniques. Health Technol.

Moorthy, R., Udupa, S., Bhagavath, S. A., & Rao, V. (2020). Centralized and automated healthcare systems: A essential smart application post Covid-19. In *3rd International Conference on Intelligent Sustainable Systems (ICISS)* (pp. 131–138). https://doi.org/10.1109/ICISS49785.2020.9316103.

Rathore, V. S., Kumawat, V., Umamaheswari, B. (2020). Coronavirus (COVID-19) in India- Statistics & Facts. In *2020 Fourth World Conference on Smart Trends in Systems, Security and Sustainability* (WorldS4) (pp. 518–521). https://doi.org/10.1109/WorldS450073.2020.9210399.

Shareena, Shahid., Mahammad. (2020). Work from home during COVID-19: Employees perception and experiences. 9.

Stogner, J., Miller, B. L., & McLean, K. (2020). Police stress, mental health, and resiliency during the COVID-19 pandemic. *American Journal of Criminal Justice, 45*. https://doi.org/10.1007/s12103-020-09548-y.

Tang, K. (2020). Risk factors and indicators for COVID-19 severity: Clinical severe cases and their implications to prevention and treatment. In *International conference on public health and data science (ICPHDS)* (pp. 333–337). https://doi.org/10.1109/ICPHDS51617.2020.00072.

Wang, R. Y., Guo, T. Q., Li, L. G., Jiao, J. Y., & Wang, L. Y. (2020). Predictions of COVID-19 infection severity based on co-associations between the SNPs of co-morbid diseases and COVID-19 through machine learning of genetic data. In *IEEE 8th International Conference on Computer Science and Network Technology (ICCSNT)* (pp. 92–96). https://doi.org/10.1109/ICCSNT50940.2020.9304990.

Tai, Y., Gao, B., Li, Q., Yu, Z., Zhu, C., & Chang, V. (2021). Trustworthy and intelligent COVID-19 diagnostic IoMT through XR and deep learning-based clinic data access. *IEEE Internet of Things Journal*, 1–1. https://doi.org/10.1109/JIOT.2021.3055804.

Covid-19 Face Mask Detection Using Deep Learning Techniques

T. P. Pushpavathi and Siranjeevi Narasimman

Abstract Covid-19 Novel Coronavirus is the most hazardous virus, and its impact affects every countries economy. It affected and killed four million+ people around the world. Still, we didn't find a medicine to cure this COVID. Even though we can't cure it, we can reduce the spreading. So, I motivated myself to reduce these covid deaths and affections. Then I decided to do a Face Mask Detection system which is the foremost reason for this pandemic. The scope of our present investigation on Face Mask Detection can be performed around the entire campus present anywhere in the world. The main motive is to detect people's face masks and to reduce this covid. So, it can be performed anywhere in the world. My methodology is to collect the dataset efficiently. Then to detect the Face Mask by processing the dataset through deep learning techniques. I conclude that I had come up with a system that detects people's Face Mask. And I conclude that this system detects people's face mask with 95% accuracy. And I believe that Covid may reduce by this system.

Keywords Face mask detection · Deep learning techniques

1 Introduction

In December 2019 was the primary instance of novel Covid (Coronavirus) contamination happened in Wuhan, China. The infection spreads principally between individuals who are in close contact with one another, commonly inside 1 m (short-range). It started spreading all over the world in a very less span of time. Due to this rapid spreading of the Coronavirus, which had declared COVID-19 as a world pandemic. For the last six months, more than 5 million cases have been infected by COVID-19 across the World. So, we should be aware and follow the safety measures properly. An individual can be contaminated when sprayers or drops containing the infection are breathed in or come straightforwardly into contact with the eyes, nose, or mouth.

T. P. Pushpavathi · S. Narasimman (✉)
Department of CSE, MSRUAS, Bangalore, India
e-mail: siranjeevinarasimman@gmail.com

V. Chang et al. (eds.), *Artificial Intelligence and Machine Learning Methods in COVID-19 and Related Health Diseases*, Studies in Computational Intelligence 1023,
https://doi.org/10.1007/978-3-031-04597-4_4

Symptoms include fever, cough, chills, loss of hunger and in cut-off conditions, respiratory issues like chest substantialness, dry sputum isolation brings about trouble in breathing lastly, and deadly may happen. On March 11, 2020, (WHO) proclaimed it as a destructive infection that acquired its underlying foundations across the globe and seriously influenced 114 nations. It killed four million+ people around the world. Scientists proved that Wearing a face veil has turned into a viable way to keep away from contamination with Coronavirus. Things such as respiratory contamination infections, poisonous and destructive gases, and the residue suspended noticeable all around can enter people's lungs as they inhale and afterward cause pneumonia, nerve harm, and poisonous responses. Specifically, the new (COVID-2019) has spread worldwide since the finish of 2019, which extraordinarily affects the security of lives and property of every person. At the point when individuals are presented with poisonous or unsafe gases, wearing covers can viably shield them from being imperiled, accordingly decreasing unnecessary misfortunes (Alankar et al., 2021). Subsequently, it is of incredibly reasonable importance to understand the veil-wearing location calculation. As of now, in places where covers should be worn (like networks, grounds, general stores, medical clinics, processing plants, stations, and so forth), the wearing of veils is typically checked physically. Nonetheless, this strategy would cause misuse of HR and low proficiency, and above all, there are issues like absent and bogus discovery. Item location innovation empowers us to utilize the camera and PC incorporated way to understand the face cover wearing location, so the reason for non-contact programmed identification is accomplished. Facial veil location should be possible utilizing semantic division. In this, the semantic division is utilized in the location of the face veil. At the point when the picture is given, on that picture, the semantic division is performed and from the semantic division, the face veil is distinguished (Singhm et al., 2021). Face veil acknowledgment that catches the facial picture. This framework utilizes the engineering provisions of the established network for face acknowledgment. Profound learning procedures are applied to develop a classifier that will gather a picture of an individual wearing a face cover and no covers. Our COVID-19 face veil discovery dataset Fig. 2: A face cover location dataset comprises "with cover" and "without cover" pictures. We will utilize the dataset to assemble a COVID-19 face veil finder with PC vision and profound getting the hang of utilizing Python, OpenCV, and TensorFlow/Keras. **Motivation—**Covid-19 Novel Coronavirus is the most hazardous virus, and its impact affects every countries economy. It affected and killed four million+ people around the world. Still, we didn't find a medicine to cure this COVID. Even though we can't cure it, we can reduce the spreading. So, I motivated myself to reduce these covid deaths and affections. Then I decided to do a Face Veil Location framework which is the chief justification for this pandemic. **Significance of Face Veil Recognition—**Face cover identification plays a vital role in reducing the spreading of COVID-19. So that people can be aware, and they would wear a mask and this COVID can be reduced. Surviving is also a basic essential, and work was made compulsory in many organizations. Face mask detection system is essential in this huge society. **Advantages of Face Mask Detection—**Face mask detection is much essential in the current scenario as it can stop spreading the pandemic efficiently. The techniques used in this method can recognize the person and detect whether they wear masks or not. We can recognize Pre preparing pictures and furthermore distinguish through live video. On the off chance that individuals wear veils, it will allow them. If not, it will give the ringer to wear a cover to keep them from infection transmission. By this COVID-19 DETECTION SYSTEM, we can reduce COVID infections by at least 50%. **Aim of the Project**—The aim of the project is to detect people, whether they are wearing masks or not, using Deep Learning Techniques. The main objective is to enable face mask detection so that people will get aware and this COVID can be reduced.

2 Related Work

In this part, a synopsis is discussed depending on the model chosen for executing this work.

In building AI classifiers to identify objects, customary strategies were utilized to remove the provisions. The Viola-Jones identifier is the most well-known, which utilizes Haar highlights with an essential picture technique to identify faces. For human location, a valuable element extractor called histogram of arranged inclinations (HOG) is a component descriptor where registers the bearings and extents of situated angles over picture cells, scale-invariant element change (SIFT) is exceptionally helpful by and by for picture coordinating and object acknowledgment, accelerate powerful component (SURF), And other, and so forth. However, profound learning-based indicators exhibited fantastic execution in the last few years because of their high component extraction capacity. For example, CNN, which assumes a fundamental part in design acknowledgment undertakings related to PC vision, is an unrivaled spatial component extraction capacity. CNN utilizes convolution parts to convolve with the unique pictures or component guides to separate more significant level elements. CNN has been taken advantage of in face acknowledgment and transcribed person acknowledgment. We proposed a CNN model for vigorous face veil location.

Problems Exist in Detection:-

There are two central issues of face cover wearing identification. One is to find the position of the face in the picture; the other is to recognize whether the face given by the informational index is wearing a cover and if the veil is worn accurately. Issues of the current item identification calculation can be credited to confront impediment, variable face scale, lopsided enlightenment, thickness, and so on. These issues genuinely influence the presentation of the calculation. Furthermore, the conventional item recognition calculation embraces the specific pursuit technique in highlight extraction, prompting issues like helpless speculation capacity, excess data, low precision and helpless continuous execution.

Loey (2021) "**A hybrid deep transfer learning model with machine learning methods for face mask detection in the era of COVID-19 pandemic**". The creators fostered another facemask-wearing condition distinguishing proof strategy. They had the alternative to describe three groupings of facemask-wearing conditions. The orders are correct facemask-wearing, erroneous facemask-wearing, and no facemask-wearing. The proposed strategy has achieved 98.70% precision in the face area stage. **Sabbir** and others have applied the Primary Part Investigation (PCA) on covered and presented face affirmation to see the person. They found that the precision of face resonation using the PCA is limited impacted by wearing covers. The affirmation precision drops to under 70% when the apparent face is covered. Furthermore, PCA was used. The makers proposed a strategy that is used for disposing of glasses from a human forward-looking facial picture. The wiped-out part was reproduced using recursive misstep compensation using PCA revamping.

In, the producers utilized the YOLOv3 calculation for face affirmation. YOLOv3 utilizes Darknet-53 as the spine. The proposed procedure accomplished 93.9% precision. It was ready on CelebA and a more extensive FACE dataset, including in excess of 600,000 pictures. The testing was the FDDB dataset. Nizam and others proposed an original GAN-based association that can normally dispose of veils covering the face district and recuperates the image by building the missing opening. The yield of the proposed model is a completed face picture that looks normal and reasonable.

The producers introduced a framework for perceiving the presence or nonappearance of a compulsory clinical cover in the working room. The overall objective is to restrict the sham positive face area as possible without missing veil recognizable pieces of proof to trigger alerts only for clinical staff who don't wear a cautious cover. The proposed system chronicled 95% precision.

Muhammad and others presented a natural method called MRGAN. The strategy depends on getting the enhancer area from the customer and using the Generative Antagonistic Organization to revamp this locale. Shaik and others used significant learning persistent face feeling portrayal and affirmation. They used VGG-16 to organize seven looks. The proposed model was ready on the KDEF dataset and achieved 88% precision. There are two focal issues of face cover wearing acknowledgment. One is to discover the circumstance of the face in the image; the other is to perceive whether the face given by the educational assortment is wearing a cover and if the veil is worn successfully. Issues of the current article identification calculation can be ascribed to confront impediment, variable face scale, lopsided brightening, thickness, and so forth. These issues genuinely influence the exhibition of the calculation. Besides, the customary article identification calculation receives the specific hunt technique in highlight extraction, prompting issues like helpless speculation capacity, repetitive data, low precision, and helpless ongoing execution.

Yu (2021) "**Face veil wearing recognition calculation dependent on further developed YOLO-v4**". A few authors have utilized the extraction of RGB shading data to perform face mask acknowledgment. Regardless, the article doesn't contemplate the occasion of non-standard wearing of veils, so the estimation's adaptability ought to be further improved. Joining YOLO-v2 and ResNet50, the makers in recognized face cover affirmation whose spine network is DarkNet-19. Regardless, DarkNet-19 has been smoothed out by CSPDarkNet53. The evacuation breakdown in our paper shows that the CSP1_X module conveys ideal results over CSPDarkNet53.

The makers raised that the blend of ResNet50 and SVM can comprehend face shroud ID and its exactness can reach up to 99.64%. In any case, the estimation takes a huge load of computational costs. Also, the blend of SSD and MobileNetV2 for cover acknowledgment was proposed in the paper. Anyway, its model plan is excessively erratic, and its display is fairly contrasted with YOLO-v4.

Just two classes are utilized in the papers referenced in the above section and the creators did not consider the impact of wearing masks sporadically on the calculation. Hence, the element extraction capacity and model practicability of these calculations should be improved. In this paper, considering further developed YOLO-v4, face cover acknowledgment is thought of and three classifications, facemask, face and WMI, are incorporated. Moreover, the element extraction capacity of this paper is improved by CSP1_X, and CSP2_X induces PANet to accelerate the course of semantic highlights and fortify component combination, hence working on the heartiness of the model.

Nagrath (2021) **"An ongoing DNN-based face cover recognition framework utilizing single shot multibox identifier and MobileNetV2"**. Face Cover disclosure models have various assortments. These can be disengaged into a couple of classes. In the Boosting-based plan, assisted falls with basic haar features were acknowledged using the Viola-Jones face locater (Viola & Jones, 2001), which was analyzed above in this part. Then a Multiview face cover finder was made impelled by the Viola-Jones locater model. What's more, a face cover finder model was settled on utilizing choice trees calculations. Face cover indicators in this classification were extremely compelling in distinguishing face veils.

In Deformable Part Model-based portrayal, the plan and headings of a couple of special faces are shown using DPM. In 2006, Ramanan proposed an Arbitrary woods tree model in face cloak acknowledgment, which exactly deduces face developments and facial stances. Li et al. (2016), one of the renowned researchers, made a DPM-based face cover marker using around 30, 000 faces isolated into covers and without covers class. His work achieved a wonderful precision of 97.14%. Further models of face cover identifiers were made by Cao et al. (2014). Normally, DPM-based face cover distinguishing proof models can achieve incredible precisions, yet they may be merciful from the very rising above the cost of estimation as a result of the usage of DPM.

In Convolutional Neural Organization-based portrayal, face locater models acquire directly from the customer's data and a short time later apply a couple of significant learning estimations on it (Girshick et al. 2015). In the year 2007, Brandt et al. (2015) considered Course CNN (2015), Yang and others considered features gathering of faces in the face acknowledgment model. Ojala and others (2002) overhauled the AlexNet design for tweaking the image dataset in extra assessment works. For uninhibited conditions, Zhu and others.

Nagrath et al. (2022) proposes a Logical Multi-Scale Area-based Convolutional Neural Organization (CMS-RCNN), which invited a basic impact on the face area models. To restrict the screw up on the substitute layers of CNN layers and deal with the uneven checks created in the veil area models. Opitz and others (2016) organized a grid incident layer. As advancement advanced, further CNN-based 3D models terminated coming up; one was proposed by Li and others (2015). It was a learning structure for face cover disclosure models. A couple of various works were done in the circle of stance affirmation, sex appraisal, limitation of places of interest, etc. The Face cover disclosure model named SSDMNV2 has been made using significant neural association modules from OpenCV and TensorFlow, which contains a Solitary Shot Multibox Identifier object recognizable proof model. Typical portrayal plans like ResNet-10, which is used as a spine design for this model and picture gathering and changed MobileNetV2 classifier has been used, MobileNetV2 classifier has been an improvement over MobileNetV1 plan classifier as it included 3 × 3 convolutional layers as the hidden layer, which is followed by various occasions the previous structure blocks. Curiously, MobileNetV2 configuration involves 17, 3 × 3 convolutional layers straight joined by a 1 × 1 convolution, a typical layer of max pooling, and a layer of the game plan. The excess affiliation is one more extension in the MobileNetV2 classifier.

Jian Cao (2020) "**Robust Deep Learning Method to Detect Face Masks**". A powerful and lively article acknowledgment estimation to therefore recognize the appearances with covers or without veils, making the plague contravention work smarter. We assembled a wide database of 9886 pictures of people with and without face covers and truly named them, then used multi-scale getting ready and picture botch systems to additionally foster YOLOv3, an article ID estimation, to normally perceive whether a face is wearing a cloak. Our preliminary outcomes show that the mean Normal Exactness (Guide) of the further created YOLOv3 computation model came to 86.3%. This work can reasonably and thusly recognize whether people are wearing covers, which diminishes the squeezing element of passing on HR for checking covers transparently and has high sober-minded application regard.

Yangro Huang (2020) "**Veiled Face Acknowledgment with Dormant Part Identification**". Considering the significant CNN, a multi-stage shroud learning framework is proposed to discover and dispose of tainted element components from acknowledgment, which sets up a cover word reference dependent on blended face pictures with arbitrary halfway impediments. **Yin** and others propose a spatial actuation variety misfortune to learn more organized face portrayals and an element enactment variety misfortune to push the interpretable depictions to be discriminative and incredible to hindrances. Additionally, a couple of procedures are proposed to recover clean faces from the impeded ones. For example, the inadequate depiction-based game plan (SRC) reproduces a solid obstruction face using a straight mix of pictures from the arrangement set alongside a pitiful limit term. In any case, SRC requires that all of the faces are from a pre-portrayed shut set, which misfires on open-set face affirmation. Zhao and others propose the RLA model, which involves two LSTM sections, intending to encode obstacle vivacious face and restore the hindered part separately. Daniel and others power the convolutional neural association to take in discriminative features from all the face regions even more by covering the specific region to which the association gives the most thought. Naturally, OFR and MFR have a tantamount task, yet the instruments of these two kinds of systems are essentially extraordinary: in OFR, the space of hindrance is uncertain, in this way, most OFR methods plan to learn features fiery to the obstruction which might appear wherever on the image; in spite of what may be generally anticipated, the general space of obstacle (i.e., covers) is known in our MFR task and the MFR strategies can benefit with the previous data on the spot. For instance, the continuous OFR method enables the part map to start over the whole face to pardon the effect of neighborhood obstacles, while our proposed procedure expects the part map to be authorized on adjacent regions which are not obstructed by a shroud. The exploratory results in like manner display the proposed procedure is more fitting for the MFR task.

Besides, the Halfway Face Acknowledgment (PFR) task expects to facilitate fragmentary face pictures with ordinary face pictures. Earlier work relies upon the hand-make close-by descriptors. With significant learning improvement, a multi-scale region-based CNNs (MR-CNN) model is proposed to achieve incredibly traditionalist and discriminative features. He and others propose an extraordinary component organizing (DFM) procedure with a sliding disaster to address midway face pictures paying little psyche to appraise. There are two bothers while clearly using these methods on the MFR task: Initially, the PFR methodologies require the pre-described partial face pictures as data sources, which are hard to perceive or semantically portray in covered appearances.

Additionally, the PRF systems disregard the overall indications of hidden face pictures, for instance, facial structure structures. Different from these strategies, the proposed technique separates the torpid part, which is solid to cover wearing and learns the overall level and mostly level features together. Learning discriminative parts has been extensively embraced in various affirmation endeavors like individual re-ID or vehicle re-ID. These systems routinely need pre-described semantic parts and use extra named data to set up the part area or division models. Regardless, it is hard to portray which part should be perceived unequivocally and is by and large legitimate for the MSR task. In order to discover torpid parts normally, a couple of methodologies are arranged ward on thought methodology, which use an extra subnet to gain a couple of regions of income and concentrate features from related areas. Regardless, the thought maps are simply surveyed from the appearance guides and needs bearing from the space of parts. Unmistakable with these thought-based techniques, we utilize the prior doubt that the veils are continually worn at the lower half of the face as a fragile restriction to work with the inactive part learning.

3 Methods and Methodology

Novel Covid is a present worldwide pandemic where the whole world is languishing. Profound learning methods can tackle complicated issues by gaining from straightforward portrayals. As profound learning methods have the equipped for learning definite portrayal and the learning the information in an interaction where the various layers are utilized successively made the profound adapting so mainstream. Execution of existing profound learning models of face cover recognition utilizes deep learning. This includes CNN Organization Framework and contrasting their presentation and examination as displayed in Fig. 1.

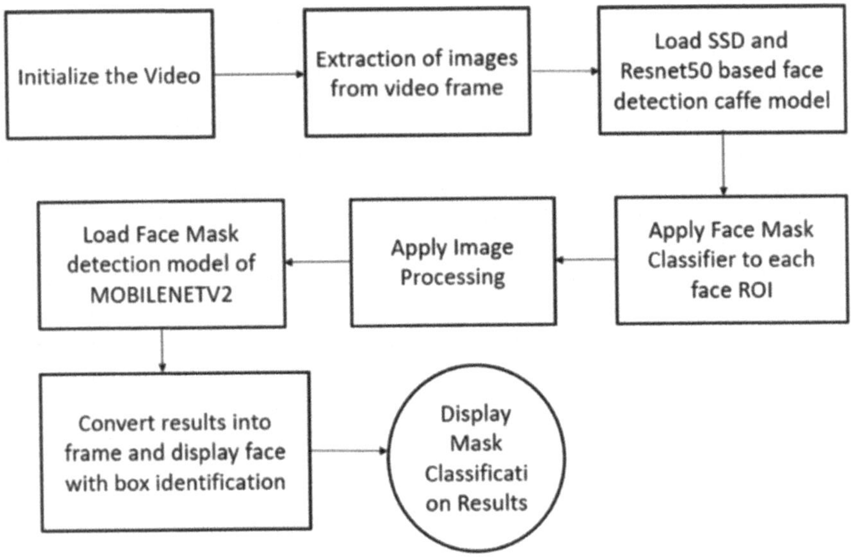

Fig. 1 Block diagram

Convolutional neural organization (CNN): CNN is a sort of Profound Neural Organization, ordinarily even minded to dissect visual symbolism. CNN is exceptionally indistinguishable from ordinary Neural Networks. It comprises neurons that are pre-prepared with regard to the information. Each neuron takes almost all inputs, speck item work is shown and accomplishes it non-directly. The whole organization shows a solitary differentiable capacity—crude picture pixels on one highlight score of the class at the other. Misfortune work like SVM/SoftMax are used and at last, a completely associated thick layer is applied (Fig. 2).

4 Dataset Description

The dataset was assembled from different open-source web applications. For example, Kaggle and GitHub 1200 picture Dataset were gathered for Veiled and Exposed Pictures. This examination directed its tests on three unique datasets. The first dataset is Real-World Masked Face Dataset. The creator made one of the greatest veiled face datasets utilized in this examination. The dataset comprises 5000 concealed countenances and 90,000 exposed appearances. It delineates tests of countenances with and without veils. In this examination, 5000 pictures for faces with veils and without veils have been utilized with an aggregate of 10,000 pictures to adjust the dataset. The dataset was utilized for the preparation, approval, and testing stages. The second dataset is a Simulated Masked Face Dataset. The dataset comprises 1570 pictures, 785 for reenacted concealed faces, 785 for exposed countenances. Models

Fig. 2 **a** Unmasked images. **b** Masked images

for pictures of the pieces of information are introduced. The dataset was utilized for the preparation, approval, what's more, trying stages. The Third dataset utilized in this examination is the Labeled Faces in the Wild. It is a reenacted covered face dataset that contains 13,000 veiled appearances for superstars around the round. It represents tests of pictures. The dataset utilized for the testing stage was a benchmark testing dataset which the proposed model won't ever prepare on it (Fig. 2).

4.1 CNN—Convolutional Neural Network

This layer is the basic square of the Convolutional Neural Organization. The term convolution infers a numerical mix of two capacities to get the third capacity. It chips away at a sliding window instrument, which helps in extricating highlights from a picture. This makes a difference in age highlight maps. CNN is explicitly intended to deal with input pictures. Their engineering is then more explicit: it is made out of two primary squares. The principal block makes the distinction of this kind of neural organization since it capacities as an element extractor.

This design is a Convolutional Neural Organization sort of engineering as displayed in above Fig. 3. I have taken this picture from the Web. There are two standard parts to a CNN designing 1. A convolution gadget that disconnects and recognizes the various arrangements of the image for examination in a collaboration called Component Extraction 2. A related layer that utilizes the yield from the convolution communication and predicts the class of the image subject to the parts isolated in past stages.

Pooling Layer:-

Applying the pooling activities can make computations go quicker by permitting a decrease in the size of the info network without losing numerous highlights. Diverse sort of pooling activities can be applied out of which some are clarified underneath:

1. Max Pooling: It takes the greatest worth present in the chosen district where the piece is presently at as the incentive for the yield of network an incentive for that cell.
2. Average Pooling: It takes the normal of the multitude of qualities that are right now in the locale where the piece is at and takes this worth Dataset Visualization. P. Nagrath. Practical Cities and Society 66 (2021) 102692 As the yield for the lattice worth of that cell shows the normal pooling activity.

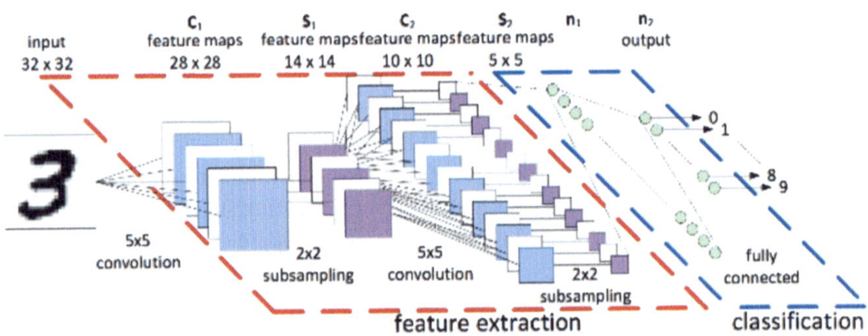

Fig. 3 CNN architecture

Dropout Layer:-

This lessens the overfitting, which might happen while preparing by dropping irregular one-sided neurons from the model. These neurons can be a piece of stowed away layers, just as noticeable layers. The probability for a neuron to be dropped can be changed by changing the dropout proportion.

Non – Linear Layer:-

These layers normally follow the convolutional layers. Most ordinarily utilized non-straight capacities incorporate various types of Rectified Linear Unit (ReLU) (Hahnloser et al. 2000), I. e., Leaky ReLU (Zhang et al. 2017), Noisy ReLU (Gulcehre et al. 2016), Exponential ReLU (Opschoor et al. 2020), and so on, sigmoid capacity just as tanh capacities. Various types of Non-Linear Functions and their conditions are shown underneath.

$$Sigmoid : \sigma(x) = \frac{1}{1 + e^{-x}}$$

$$Leaky\ \mathrm{Re}lu = f(x) = \max(0.1x, x)$$

$$\mathrm{Tanh} : f(x) = \tanh(x)$$

$$Maxout : f(x) - \max(w_1^T x + b_1, w_2^T x + b_2)$$

$$\mathrm{Re}lu : f(x) = \max(0, x)$$

$$ELU : f(x) = \begin{cases} xx \geq 0 \\ \alpha(e^x - 1)x < 0 \end{cases}$$

Fully Connected Layer:-

These layers are appended in the model and have full associations with actuations layers. These layers help in grouping the given pictures in multi-class or paired characterization. SoftMax is an illustration of actuation capacities utilized in these layers, and it gives the consequence of anticipated yield classes as far as likelihood.

Linear Bottlenecks:-

As different grid augmentations can't be decreased to a solitary mathematical activity, non-direct enactment capacities, as are ReLU6 applied in the neural organizations so that a few errors can be without any problem taken out. A multi-facet neural organization can be assembled through this since the ReLU actuation work leaves the qualities under 0. The elements of the organization increment by heightening the quantity of channels to challenge the deficiency of data. For a switched remaining square, layers of the squares are compacted. Furthermore, the opposite technique is done as done previously. This is done at the point where the skip associations are connected; this could influence the execution of the organization. In order to manage this, the idea of the direct bottleneck was presented. Prior to adding the square to introductory actuation, the last convolution of the left-over block is given a direct yield.

Mask Detection Model:-

For the reasons for momentum research, pictures having a place with the "veiled face" and "exposed face" classes were chosen from the information base. While examining the between class dissemination, an unevenness of classes was uncovered. The majority of the pictures have a place with the "veiled face" class (6769 versus 2086). This issue has an adverse consequence on AI calculations (Ryumina & Karpov, 2020b). To address it, we looked for face pictures "exposed face" utilizing the Google web crawler and the Web until the classes were even. Since the outline of the proposed technique for facial cover identification. The International Archives of the Photogrammetry, Remote Sensing and Spatial Information Sciences, Volume XLIV-2/W1-2021 fourth Int. Worksh. on "Photogrammetric and PC vision strategies for video reconnaissance, biometrics and biomedicine", 26–28 April 2021, Moscow, Russia This commitment has been peer-checked. https://doi.org/10.5194/isprs-doc uments XLIV-2-W1-2021-177-2021 | © Author(s) 2021. CC BY 4.0 License. 178 as of now gives face region arranges (bouncing boxes), face discovery for new examples was performed utilizing the SSD. The SSD is carried out in the Deep Neural Networks (DNN) module of the Open-Source Computer Vision Library (OpenCV, 2021, Pulli, 2012). Then, at that point, the recognized regions have physically kept an eye on the rightness of regions with faces. In this way, the Clinical Mask Extended Dataset was gathered and explained.

Proposed Model:-

The presented model incorporates two primary parts, the first part is profound moving learning (ResNet50) as a component extractor. Furthermore, the subsequent part is an old-style AI-like choice trees and gathering. ResNet-50 has accomplished better outcomes when it is utilized as an element extractor. The Figure outlines the proposed old-style move learning model. Essentially, the ResNet50 is utilized for the element extraction stage, while the conventional AI model is utilized in the preparation, approval, and testing stage. A remaining neural organization (ResNet) is a sort of profound exchange learning in light of remaining learning. A wide range of ResNet-101, ResNet-50, and ResNet-18 are adaptations of ResNet to dispose of the issue of disappearing slopes that have their particular remaining square. ResNet-50 with 50-layers are profound, start with a convolution layer, and end with a fully connected layer, and in the middle followed by 16 remaining bottlenecks obstructs each square has three layers of convolution layer as displayed in Figure. In the arrangement, the last layer in ResNet-50 was eliminated and supplanted with three customary AI classifiers, choice tree, and gathering to work on our model execution. The primary commitment of this examination is to build a model, choice trees, and outfit that don't overfit the preparing measure.

5 Result and Analysis

This part clarifies the consequences of pre-preparing the information and afterward preparing the information. Initial, a capacity named sorted_ alphanumerically is used to sort the rundown in lexicographical requests. A capacity pre-preparing is characterized, which takes the envelope to the dataset as info, then, at that point, stacks all the records from the envelope and resizes the pictures as indicated by the model. Then, at that point, after the rundown is arranged, the pictures are changed over into tensors. Then, at that point, the rundown is changed over to the NumPy exhibit for quicker estimation. After applying pre-processing, the exactness of our model expanded altogether. After this, the course of information expansion is applied to build the exactness subsequent to preparing the model. The Relative examination for Coronavirus Face Veil acknowledgment through Profound Learning models is executed and the outcomes are as displayed independently. Dataset of 1200 pictures containing Covered and Exposed pictures. The precision of a dataset model is characterized and estimated utilizing certain boundaries, which are as per the following. The Preparation and Approval Misfortune and Exactness of Covered and Exposed Dataset of Ages is 25 (Figs. 4 and 5).

5.1 Comparative Analysis

The execution for Face Veil acknowledgment of Coronavirus utilizing different profound models is as displayed in above Table 1. Contrasting the above table shows a similar examination of face cover and exposed dataset with precision, large scale and weighted avg.

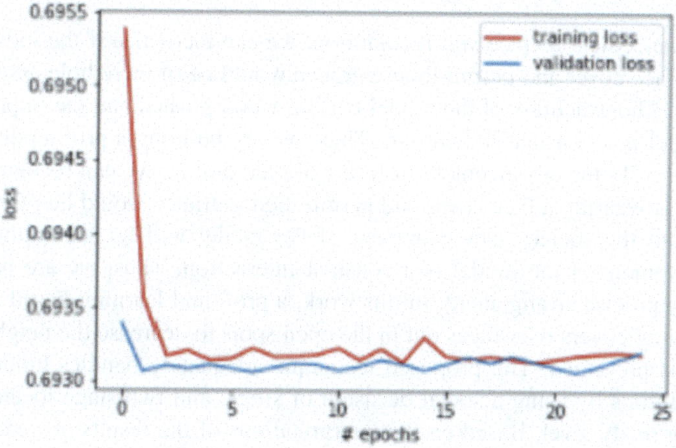

Fig. 4 Graph for training and validation loss of masked and unmasked dataset

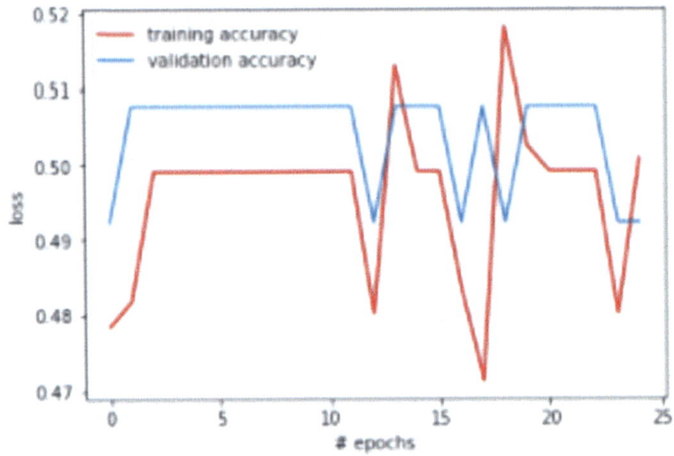

Fig. 5 Graph for training and validation accuracy of masked and unmasked dataset

Table 1 Comparative analysis

	Precision	Recall	F1-Score	Support
With mask	0.99	0.99	0.99	433
Without mask	0.99	0.99	0.99	386
Accuracy			0.99	819
Macro avg	0.99	0.99	0.99	819
Weighted avg	0.99	0.99	0.99	819

6 Conclusion

By the advancement of face veil recognition, we can recognize if the individual is wearing a face cover and permit their entrance would be of incredible assistance to the society. The exactness of the model will be accomplished and the improvement of the model is a constant interaction. Thus, we are building a profoundly precise arrangement. By the advancement of face veil recognition, we can recognize if the individual is wearing a face cover and permit their entrance would be of incredible assistance to the society. The exactness of the model will be accomplished and the improvement of the model is a constant interaction. Thus, we are building a profoundly precise arrangement. In this work, a profound learning-based approach for perceiving covers over faces out in the open spots to decrease the neighborhood of Covid is presented. The proposed technique adequately handles hindrances in thick conditions by using a social occasion of single and two-stage locaters at the pre-getting ready level. Based on the interpretations of the results presented in the previous chapters, I conclude that this "Covid-19 Face Mask Detection system" has achieved 95% accuracy. **Suggestions for future directions**—In this pandemic circumstance, where the entire world is dreaming of getting back to typical daily practice, this framework will assume a successful part in checking the utilization of face covers in work environments. By advancing this framework, we can identify the veil all over and permit his entrance into the work environment. This framework likewise adds to public medical care, as it helps in keeping the climate solid. This framework can be appropriately utilized in open regions with inserted frameworks for application in air terminals, rail line stations, workplaces, schools, and public spots to guarantee that public wellbeing rules are followed. At long last, the work opens fascinating future headings for scientists. Initially, the proposed procedure can be coordinated into any high-goal video observation gadgets and not restricted to cover location as it were. Also, the model can be reached out to recognize facial tourist spots with a facemask for biometric purposes.

References

Abusham, E. E. A., Jin, A. T. B., Kiong, W. E., & Debashis, G. Face recognition based on non-linear feature approach. Retrieved from http://citeseerx.ist.psu.edu/viewdoc/summary?doi=10.1.1.127.1596.

Alankar, B., Kaur, H., Ahsaan, S., Sharma, G., & Chang, V. (2021). Towards reviewing an immediate impact of COVID-19 on the integrative world economy: An evolving perspective. *Journal of Global Information Management (JGIM), 30*(4), 1–19.

A scalable framework for smart COVID surveillance in the workplace using Deep Neural Networks and cloud computing. Expert Systems (2021). e12704. Retrieved from https://onlinelibrary.wiley.com/doi/full/10.1111/exsy.12704.

Bhadani, A. K., & Sinha, A. A facemask detector using machine learning and image processing techniques. *Engineering Science and Technology, An International Journal*, 1–8.

Cao, X., Wei, Y., Wen, F., & Sun, J. (2014). Face alignment by explicit shape regression. In *International journal of computer vision, 107*(2), 177–190.

Chavda, A., Dsouza, J., Badgujar, S., & Damani, A. (2021). Multi-stage cnn architecture for face mask detection. In *2021 6th International Conference for Convergence in Technology (I2CT)* (pp. 1–8). https://ieeexplore.ieee.org/abstract/document/9418207.

Ding, F., Peng, P., Huang, Y., Geng, M., & Tian, Y. Masked face recognition with latent part detection. In *Proceedings of the 28th ACM International Conference on Multimedia* (pp. 2281–2289). Retrieved from https://dl.acm.org/doi/abs/10.1145/3394171.3413731.

Ge, S., Li, J., Ye, Q., & Luo, Z. (2017). Detecting masked faces in the wild with lle-cnns - Proceedings of the IEEE conference on computer. Retrieved from https://openaccess.thecvf.com/content_cvpr_2017/html/Ge_Detecting_Masked_Faces_CVPR_2017_paper.html.

Gulcehre, C., Moczulski, M., Denil, M., & Bengio, Y. (2016). Noisy activation functions. In *International conference on machine learning* (pp. 3059–3068). PMLR.

Heo, J., Abiantun, R., Xie, C., & Kumar, B. V. K. V. (2006). Class dependent kernel discrete cosine transform features for enhanced holistic face recognition in FRGC-IIM Savvides. In *2006 IEEE International Conference on Acoustics Speech and Signal Processing*. Retrieved from https://ieeexplore.ieee.org/abstract/document/1660310.

Hsu, R. L., Abdel-Mottaleb, M., & Jain, A. K. (2002). Face detection in color images. IEEE transactions on pattern analysis and machine intelligence, *24* (5), 696–706. Retrieved from https://ieeexplore.ieee.org/abstract/document/1000242.

Hussain, G. K. J., &, Natarajan, S. Subpixel based image scaling using continuous domain analysis.

IEEE Transactions on Industrial Informatics EDITORIAL: Advanced Deep Learning Techniques for COVID-19. (2021). *IEEE Transactions on Industrial Informatics, 17*(9), 6476–6478. Retrieved from https://pesquisa.bvsalud.org/global-literature-on-novel-coronavirus-2019-ncov/resource/pt/covidwho-1307657.

Jiang, M., Fan, X., & Yan, H. (2005). Retinamask: A face mask detector. arXiv preprint. https://arxiv.org/abs/2005.03950.

Kumar, A., Kaur, A., & Kumar, M. (2018). Face detection techniques: a review. *Artificial Intelligence Review, 52*(2), 927–948. Retrieved from https://link.springer.com/article/10.1007/s10462-018-9650-2.

Kumari, S., Ranjith, E., Gujjar, A., Narasimman, S., & Zeelani, H. S. A. S. (2021). Comparative analysis of deep learning models for COVID-19 detection. *Global Transitions Proceedings*. https://doi.org/10.1016/j.gltp.2021.08.030.

Loey, M., Manogaran, G., Taha, M. H. N., & Khalifa, N. E. M. (2021). A hybrid deep transfer learning model with machine learning methods for face mask detection in the era of the COVID-19 pandemic. *Measurement, 167*, 108288. https://doi.org/10.1016/j.measurement.2020.108288.

Meenpal, T., Balakrishnan, A., & Verma, A. (2019). Facial mask detection using semantic segmentation. In *2019 4th International Conference on Computing, Communications and Security*. Retrieved from https://ieeexplore.ieee.org/abstract/document/8888092.

Meena, D., & Sharan, R. (2016). An approach to face detection and recognition. In *2016 International Conference on Recent Advances and Innovations*. Retrieved from https://ieeexplore.ieee.org/abstract/document/7939462.

National Collaborating Centre for Women's and Children's Health. (UK RCOG Press). Surgical site infection: prevention and treatment of surgical site infection. Retrieved from https://pubmed.ncbi.nlm.nih.gov/21698848/.

Nagrath, P., Jain, R., Madan, A., Arora, R., Kataria, P., & Hemanth, J. (2021). SSDMNV2: A real time DNN-based face mask detection system using single shot multibox detector and MobileNetV2. *Sustainable Cities and Society, 66*, 102692. https://doi.org/10.1016/j.scs.2020.102692.

Nagrath, P., Thakur, N., Jain, R., Saini, D., Sharma, N., & Hemanth, J. (2022). Understanding New Age of Intelligent Video Surveillance and Deeper Analysis on Deep Learning Techniques for Object Tracking. In IoT for Sustainable Smart Cities and Society (pp. 31–63). Springer, Cham.

Opschoor, J. A., Petersen, P. C., & Schwab, C. (2020). Deep ReLU networks and high-order finite element methods. *Analysis and Applications, 18*(05), 715–770.

Lee, D. H., Chen, K. L., Liou, K. H., Liu, C. L., & Liu, J. L. (2020). Deep learning and control algorithms of direct perception for autonomous driving. *Applied Intelligence, 51*(1), 237–247. Retrieved from https://link.springer.com/article/10.1007/s10489-020-01827-9.

Li, C., Cao, J., & Zhang, X. Robust deep learning method to detect face masks. In *Proceedings of the 2nd International Conference on Artificial Intelligence*. Retrieved from https://dl.acm.org/doi/abs/10.1145/3421766.3421768.

Militante, S. V., & Dionisio, N. V. (2020). Real-time facemask recognition with alarm system using deep learning. In *2020 11th IEEE Control and System Graduate Research Colloquium (ICSGRC)* (pp. 106–110). Retrieved from https://ieeexplore.ieee.org/abstract/document/9232610.

Rahman, M. M., Manik, M. M. H., Islam, M. M., Mahmud, S., & Kim, J. H. (2020). An automated system to limit COVID-19 using facial mask detection in smart city network. In *2020 IEEE International IOT, Electronics and Mechatronics Conference*. Retrieved from https://ieeexplore.ieee.org/abstract/document/9216386.

Skinner, M. W., & Sutton, B. A. (2019). Do anaesthetists need to wear surgical masks in the operating theatre? A literature review with evidence-based recommendations *Anaesthesia and Intensive Care, 29* (4), 331–338. Retrieved from https://journals.sagepub.com/doi/abs/10.1177/0310057X0102900402.

Sun, W., Song, Y., Chen, C., Huang, J., & Kot, A. C. (2020). Face spoofing detection based on local ternary label supervision in fully convolutional networks. *IEEE Transactions on Information Forensics and Security, 15*, 3181–3196. Retrieved from https://ieeexplore.ieee.org/abstract/document/9056824.

Singh, A., Jindal, V., Sandhu, R., & Chang, V. (2021). A scalable framework for smart COVID surveillance in the workplace using deep neural networks and cloud computing. *Expert Systems*, e12704.

Viola, P., & Jones, M. (2001). Rapid object detection using a boosted cascade of simple features. In *Proceedings of the 2001 IEEE computer society conference on computer vision and pattern recognition. CVPR 2001*(Vol. 1, pp. I-I). IEEE.

Vu, N. S. A Face recognition with patterns of oriented edge magnitudes. In *Caplier European Conference on Computer Vision* (pp. 313–326). Retrieved from https://link.springer.com/book/10.1007/978-3-642-15549-9.

Wang, Z., Wang, G., Huang, B, Xiong, Z., Hong, Q., & Wu, H. (2020). Masked face recognition dataset and application. arXiv preprint. https://arxiv.org/abs/2003.09093.

Yu, J., & Zhang, W. (2021). Face mask wearing detection algorithm based on improved YOLO-v4. *Sensors, 21* (9), 3263. Retrieved from https://www.mdpi.com/1424-8220/21/9/3263.

Zhang, X., Zou, Y., & Shi, W. (2017). Dilated convolution neural network with LeakyReLU for environmental sound classification. In 2017 *22nd international conference on digital signal processing (DSP)* (pp. 1–5). IEEE.

Zhao, Z. Q., Zheng, P., Xu, S., & Wu, X. (2019). Object detection with deep learning: A review. *IEEE Transactions on Neural Networks and Learning Systems, 30* (11), 3212–3232. Retrieved from https://ieeexplore.ieee.org/abstract/document/8627998.

Computational Intelligence in Identification of Some FDA Approved Drug Compounds for Treatment of COVID-19

Debanjan Mitra and Pradeep K. Das Mohapatra

Abstract Due to unavailability of FDA approved drug for COVID-19, pursuing of available drugs are highlighted to stop COVID-19. *Insilico* investigation by molecular docking and molecular dynamics simulation help to identify some FDA pre-approved drugs which have a therapeutic effect on SARS-CoV-2. In this study, four drug compounds have been identified by descriptor properties, molecular docking, and molecular dynamics simulation. Between them, Darunavir appeared as the best drug molecule to inhibit the 3C like main protease of SARS-CoV-2. It showed − 9.1 kcal/mol binding energy in molecular docking with 3C like main protease of SARS-CoV-2. This study also enlightens on the theory "one molecule, multiple targets". Multiple target protein was docked by every single drug compound, to check their high therapeutic effect. Molecular dynamics simulations indicate the stable binding of drugs with the target protein. Until the approval of any drug for COVID-19, Darunavir might use as an anti-covid drug.

Keywords COVID-19 · Molecular docking · Molecular dynamics simulation · Descriptor properties · Darunavir

1 Introduction

There are many types of viruses present in the earth that creates disease in human. Currently, the deadly effect of COVID-19 hits the world very badly. Almost every country in the world, face its lethal nature. It was 1st time reported in late December 2019 in Wuhan in China. Then it makes itself a pandemic viral disease in a very short time. Still, now, 21.6cr people have been infected by SARS-CoV-2. Between them, 19.3cr people have recovered from its hand, but 45.07 lakhs lost in this battle (Mitra et al., 2020). Countries like the USA, India, Brazil, and Italy have been badly affected by the COVID-19. It is a type of coronavirus that creates severe acute respiratory syndrome (SARS). It was originated from MARS (Middle East respiratory

D. Mitra · P. K. Das Mohapatra (✉)
Department of Microbiology, Raiganj University, Raiganj, West Bengal 733134, India
e-mail: pkdmvu@gmail.com

V. Chang et al. (eds.), *Artificial Intelligence and Machine Learning Methods in COVID-19 and Related Health Diseases*, Studies in Computational Intelligence 1023,
https://doi.org/10.1007/978-3-031-04597-4_5

syndrome). SARS has 1st time reported in 2002. But it was not so much lethal like SARS-CoV-2. Only 8000 people were infected and 774 people have lost their lives in the 2002 SARS outbreak. *Insilico* investigations of SARS and SARS-CoV-2 have revealed the effect of mutations that make SARS-CoV-2 undefeatable (Mitra et al., 2020). However, the process of vaccination has been started in many countries.

SARS-CoV-2 is a positive-sense RNA virus and it is 30 kb in length. The genome encodes 4 structural and 16 non-structural essential proteins (Kim et al., 2020). TMEM189–UBE2V1 was reported as the most crucial gene locus that can increase the severity by interacting with the IL-1 signaling pathway (Wang et al., 2020). During infection, it infects pneumocytes of the lungs, later it causes fatal respiratory anguish. Genomic analysis revealed the COVID-19 is closely related to Bat-CoV (Suzuki et al., 2020; Lu et al., 2020; Paraskevis et al., 2020). STAT1, UBE2L6, and LCN2 are acting as core genes during COVID-19 infection (Li et al., 2020a). Pathak et al. (2021) have characterized 27 genes and identified some coagulation-related clinical symptoms of COVID-19. Kasela et al. (2021) have integrated genome-scale CRISPR loss-of-function screens to identify the genes which are associated with the risk factors of COVID-19. They have identified two genes, named *SLC6A20* and *CXCR6* as recognized causal genes that increase the risk. Nowadays, *insilico* investigations on protein sequences and structures of microbes are playing a crucial role to identify their structure, function, mutations and evolutions (Mitra & Mohapatra, 2021a, b). Tilocca et al. (2020) described the role of immune-proteomics in fighting COVID-19 infections. Non-structural proteins nsp9 and nsp10 have interacted with the NF-kappa-B-repressing factor (Li et al., 2020b). Mitra et al. (2020) identified a cyclic salt bridge in protease of SARS-CoV-2 that was never seen before in any viral protein. They have also identified some favorable mutations which make significant differences between SARS and SARS-CoV-2.

Researchers have already proposed many FDA-approved drugs for the treatment of COVID-19. In the beginning, hydroxychloroquine was offered for treatment of COVID-19 as it shows −7.28 kcal/mol and −6.7 kcal/mol binding energies in molecular docking with COVID-19 proteins (Amin & Abbas, 2020; Hosseini & Amanlou, 2020). At present, remdesivir is using by some countries due to the lack of FDA-approved COVID-19 drugs. It shows −7.6 kcal/mol binding energies (Elfiky, 2020). Drugs like lopinavir, ritonavir, and oseltamivir were identified after molecular docking and molecular dynamics simulations. They have shown −4.1 kcal/mol, −5.11 kcal/mol, −4.65 kcal/mol docking scores (Muralidharan et al., 2021). Lobo-Galo et al. (2021) had reported 3 chemical compounds as anti-COVID drugs. Captopril used as an anti-hypertension drug showed −4.7 kcal/mol energy in molecular docking. Diethyldithiocarbamate, disulfiram, were showed −2.7 kcal/mol, −4 kcal/mol respectively. Both drugs were used in chronic alcoholism. Anti-inflammatory drugs like carvacrol, oleanolic acid, and ursolic acid were also proposed by researchers due to their moderate binding energies. They have shown −4 kcal/mol, −6 kcal/mol, and −5.9 kcal/mol energies during molecular docking (Kumar et al., 2020). An anti-amoebic drug named emetine has been taken for investigation by *insilico* analysis using virtual screening. It was showed −6.5 kcal/mol energy (de Oliveira et al., 2021). Bioactive compounds from edible mushrooms were also showed excellent therapeutic effects in molecular docking (Mitra & Mohapatra, 2021c).

Artificial intelligence has shown its enormous role in the detection, identification, and investigation of COVID-19 (Abdel-Basset et al., 2021a). To overcome the image segmentation problem, a slime mold algorithm (SMA) was developed. The findings show that their suggested method beats SMA in terms of Kapur's entropy (Abdel-Basset et al., 2020). A semi-supervised few-shot segmentation (FSS) method was developed for the efficient segmentation of SARS-CoV-2 by lung CT scans (Abdel-Basset et al., 2021b). A unique XR and Deep Learning-based IoMT solution were developed for the COVID-19 telemedicine diagnosis, combining VR/AR remote surgical plan/rehearse hardware, bespoke 5G cloud computing, and deep learning algorithms to offer real-time COVID-19 treatment scheme hints (Tai et al., 2021). Researchers have developed a deep learning-based multi-objective optimization approach for detecting infected individuals by coronavirus through X-rays. Extensive testing indicates that the proposed model can accurately identify X-ray pictures with highly precise, memory, selectivity, and F1-score rates (Dhiman et al., 2021).

In this investigation, four FDA-approved drugs have been identified from an online drug bank based on molecular docking and molecular dynamics simulations. Plerixafor act as immunostimulant in cancer patients (Bapatu et al., 2015). Palbociclib is using in the treatment of breast cancer (Finn et al., 2016). Darunavir and Dolutegravir are one type of anti-viral drugs which have usually use in the treatment of HIV (Capetti et al., 2017). Binding energies between ligand and target protein have revealed their nature as a drug.

2 Material and Methods

2.1 Retrieve of Drug Compounds and Target Proteins

All drug compounds i.e., Plerixafor, Darunavir, Dolutegravir and Palbociclib (Fig. 1) were extracted from the largest drug database Pubchem (Kim et al., 2016). All files were downloaded in.sdf format. All ligands were optimized by the semi-empirical process using PM3 (Braga et al., 2004) by VEGAZZ (Pedretti et al., 2021) software. Target proteins were retrieved from the RCSB PDB database (Kouranov et al., 2006). Nsp9 RNA-replicase (6WXD), nucleocapsid protein (7C22), 3C like main protease (7BRO) were taken as targets based on previous research (Mitra & Bose, 2021).

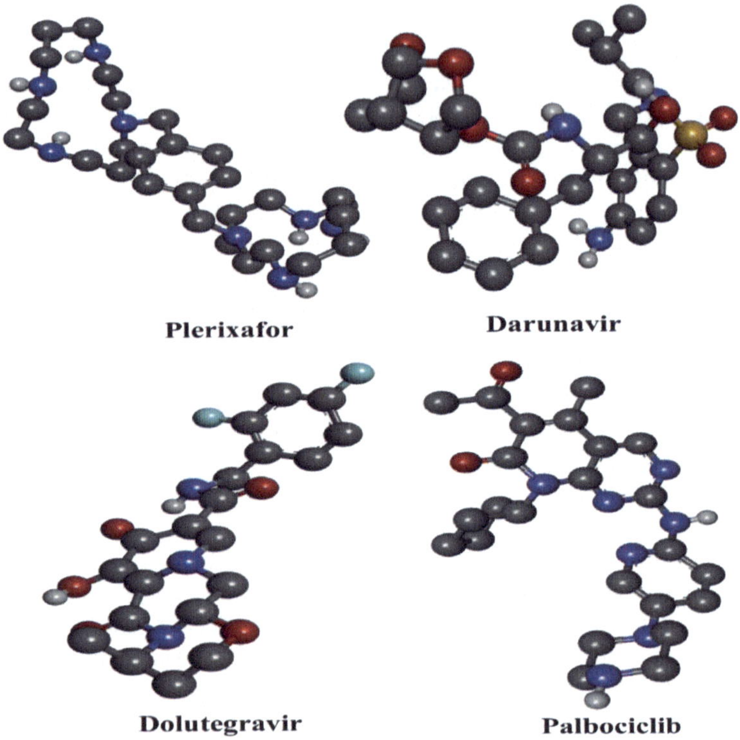

Fig. 1 3D structure of Plerixafor, Darunavir, Dolutegravir and Palbociclib

2.2 Drug Likeness Properties

Drug The Drug-likeness of Plerixafor, Darunavir, Dolutegravir, and Palbociclib has been evaluated by the Lipinski rule of five (Lipinski et al., 1997). This rule can describe the safe use and toxicity of the ligand compounds. This rule applied to oral drugs only. A slight deviation of parameters can indicate that the lead molecule will use as an injectable drug (Zhang & Wilkinson, 2007).

2.3 Molecular Docking and Dynamics

For molecular docking, Autodock 4.2 (Rizvi et al., 2013) of the Scripps Research Institute was used. The ligand has also prepared by this software. All rotatable bonds were frozen and non-polar hydrogens were added (Mitra et al., 2021a). With correct dimensions and space of grids for docking were chosen based on previous research (Mitra et al., 2021b). A very small (10 ns) molecular dynamics simulations were performed by CABS-flex 2.0 server (Kuriata et al., 2018). Figures were represented by BIOVIA Discovery Studio Visualizer (Design, 2014).

3 Results and Discussions

3.1 Drug-Likeness Properties

Drug-likeness properties were evaluated by Lipinski's rule of five (Lipinski et al., 1997). Octanol-water partition coefficient or LogP of a drug compound always will be smaller than +5, molecular weight (MW) of drug compounds will be $\leq$500, hydrogen bond acceptor (HBA) of the lead molecule will be $\leq$10, hydrogen bond donor (HBD) of drug compound will be $\leq$5. In this investigation, we have also calculated some additional principal descriptor properties like total polar surface area (TPSA), volume, Davies' HLB, and ovality.

All the drug compounds almost fulfil the Lipinski rule of five (Table 1). So those drugs can be used as oral drugs. The molecular weight (MW) of Darunavir is higher than other drug compounds. The molecular weight of Darunavir and Plerixafor is higher than 500. The number of hydrogen bond donors is slightly higher in Plerixafor. Octanol-water partition coefficient or LogP provides knowledge about the lipophilicity of a substance. Lipophilicity is a crucial character in the determination of a compound's absorption, dispersal, metabolism, and excretion in the human body (Arnott et al., 2013). The positive LogP value of Darunavir, Dolutegravir, and Palbociclib is indicating their lipophilic nature whereas the negative value of Plerixafor indicates its hydrophilic nature i.e., it has a higher affinity for the aqueous medium. Polar surface area is high in Darunavir. They can be permeable for cell membrane, but only Plerixafor can penetrate the blood–brain barrier due to its low TPSA. The volume of Plerixafor is the highest among all the drug compounds. Increasing molecular weight has also increased the ovality of all drug compounds. Davies's HLB is indicating about hydrophilic-lipophilic balance. The high value of Darunavir and Dolutegravir possess slightly the character of emulsifiers.

3.2 Molecular Docking Studies and Molecular Dynamic Simulations

Molecular docking studies between ligand and target protein revealed about the binding and interactions. Multiple targets inhibited by a single ligand can increase the effectiveness of that drug (Table 2).

Results of molecular docking showed promising binding energies in protein–ligand interactions. Plerixafor showed −8 kcal/mol energy in interaction with Nsp9

Table 1 Lipinski rule of five with some other additional descriptor properties of Plerixafor, Darunavir, Dolutegravir and Palbociclib

Name	PubChem ID	MW	HBD	HBA	LogP	TPSA	Volume	Ovality	Davies' HLB	Dipole
Plerixafor	65,015	502.8	6	8	−1.04	78.64	531.98	1.6809	7	6.4495
Darunavir	213,039	547.67	3	9	4.32	140.43	490.53	1.7506	15.2	7.3564
Dolutegravir	54,726,191	419.38	2	8	0.92	100.87	345.56	1.5729	10.2	4.2665
Palbociclib	5,330,286	447.54	2	8	2.9	105.04	410.58	1.6543	7	35.052

Table 2 Molecular docking score with ligand and target proteins i.e., Nsp9 RNA-replicase (6WXD), nucleocapsid protein (7C22), 3C like main protease (7BRO)

Name	PubChem ID	6WXD (Kcal/mol)	7C22 (Kcal/mol)	7BRO (Kcal/mol)
Plerixafor	65,015	−8	−8.7	−8.5
Darunavir	213,039	−8.7	−9.1	−9.1
Dolutegravir	54,726,191	−7.4	−8.7	−8.5
Palbociclib	5,330,286	−7.8	−8.9	−8.4

RNA-replicase (6WXD). Binding energies were increased in interaction with nucleocapsid protein (7C22) and 3C like the main protease (7BRO). Almost an equal number of Vander Waals interactions were designed by the Plerixafor. Darunavir showed the highest binding energies between all drug compounds. It showed −9.1 kcal/mol energy in interactions with nucleocapsid protein (7C22) and 3C like main protease (7BRO). Besides the formation of conventional hydrogen bonds, Darunavir formed multiple alkyl and pi-alkyl interactions to make a strong binding with target proteins. One pi-pi T-shaped interaction was also found in Darunavir and 3C like main protease of SARS-CoV-2. PHE294 of 3C like main protease has formed three interactions with Darunavir. Dolutegravir showed −7.4 kcal/mol, −8.7 kcal/mol, and −8.5 kcal/mol binding energies with interactions with proteins of SARS-CoV-2. Molecular dynamic simulations showed rugged and stable binding of Darunavir with 3C like main protease (7BRO) of SARS-CoV-2. Darunavir does not fluctuate much than the wild type of protein. Comparative study of root-mean-square fluctuation (RMSF) between Darunavir and 3C like main protease (without ligand) revealed that Darunavir might use as an anti-COVID drug due to its highly steady and well-built binding nature (Figs. 2, 3, 4, 5 and 6).

4 Conclusion

Insilico molecular docking and molecular dynamic simulations give the knowledge on drug design and drug synthesis. It also enlightens on the targets of the virus. Due to the lack of FDA-approved drugs and the high mortality rate of COVID-19, *insilico* investigation can play a vital role in the remedy of COVID-19. Some techniques which are centered on artificial intelligence are playing a crucial role in the detection of COVID-19. In this investigation, drugs are showing promising results than previously proposed drugs. These drugs are already approved by FDA for different diseases and are available in the market. Darunavir manifests the highest binding energy in molecular docking studies. Molecular dynamics simulations indicate its well-built and stable binding with the target protein. So, until any approval of drug from FDA for COVID-19, Darunavir might be used as an anti-COVID drug.

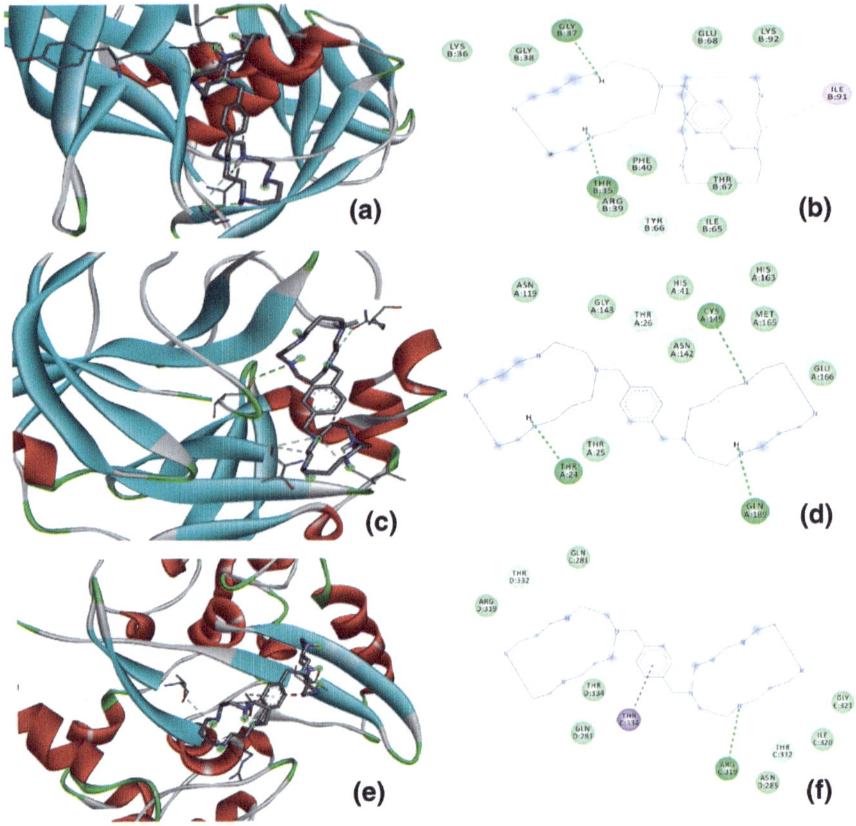

Fig. 2 **a** 3d ligand–protein interaction of Plerixafor and Nsp9 RNA-replicase (6WXD), **b** 2d ligand–protein interaction of Plerixafor and Nsp9 RNA-replicase, **c** 3d ligand–protein interaction of Plerixafor and nucleocapsid protein (7C22), **d** 2d ligand–protein interaction of Plerixafor and nucleocapsid protein (7C22), **e** 3d ligand–protein interaction of Plerixafor and 3C like main protease (7BRO), **f** 2d ligand–protein interaction of Plerixafor and 3C like main protease (7BRO)

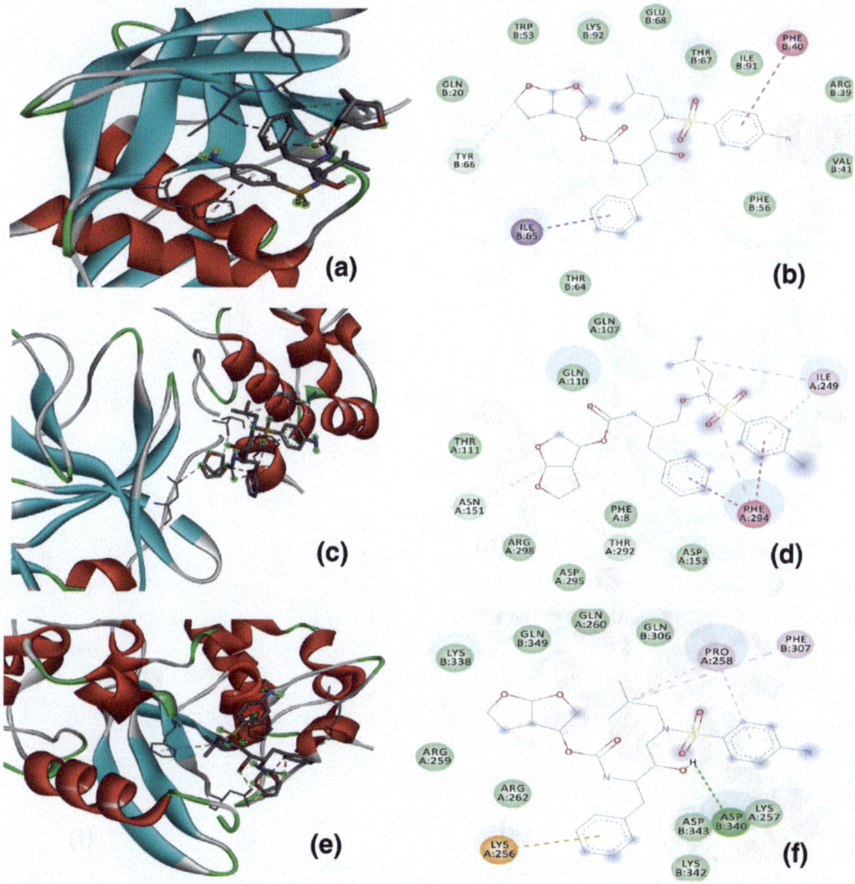

Fig. 3 **a** 3d ligand–protein interaction of Darunavir and Nsp9 RNA-replicase (6WXD), **b** 2d ligand–protein interaction of Darunavir and Nsp9 RNA-replicase, **c** 3d ligand–protein interaction of Darunavir and nucleocapsid protein (7C22), **d** 2d ligand–protein interaction of Darunavir and nucleocapsid protein (7C22), **e** 3d ligand–protein interaction of Darunavir and 3C like main protease (7BRO), **f** 2d ligand–protein interaction of Darunavir and 3C like main protease (7BRO)

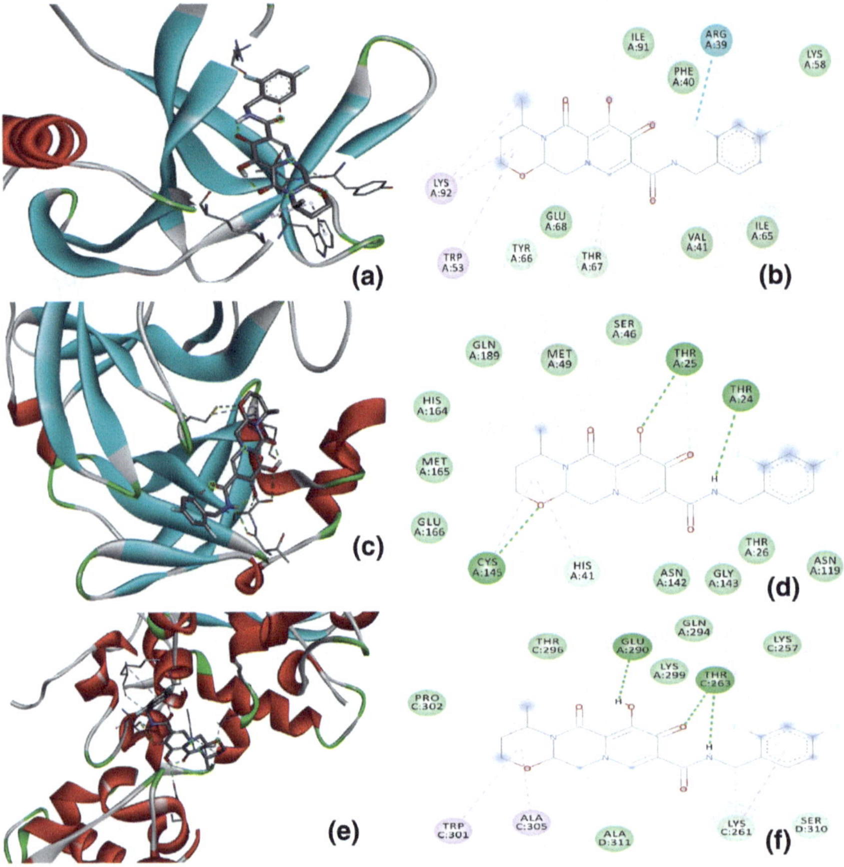

Fig. 4 **a** 3d ligand–protein interaction of Dolutegravir and Nsp9 RNA-replicase (6WXD), **b** 2d ligand–protein interaction of Dolutegravir and Nsp9 RNA-replicase, **c** 3d ligand–protein interaction of Dolutegravir and nucleocapsid protein (7C22), **d** 2d ligand–protein interaction of Dolutegravir and nucleocapsid protein (7C22), **e** 3d ligand–protein interaction of Dolutegravir and 3C like main protease (7BRO), **f** 2d ligand–protein interaction of Dolutegravir and 3C like main protease (7BRO)

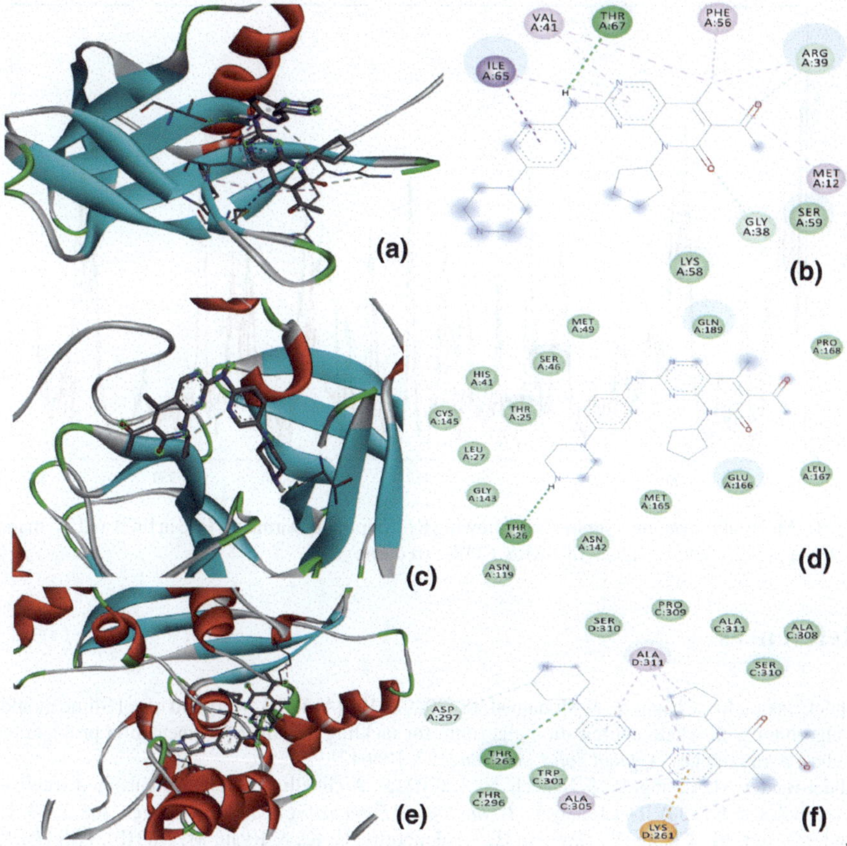

Fig. 5 **a** 3d ligand–protein interaction of Palbociclib and Nsp9 RNA-replicase (6WXD), **b** 2d ligand–protein interaction of Palbociclib and Nsp9 RNA-replicase, **c** 3d ligand–protein interaction of Palbociclib and nucleocapsid protein (7C22), **d** 2d ligand–protein interaction of Palbociclib and nucleocapsid protein (7C22), **e** 3d ligand–protein interaction of Palbociclib and 3C like main protease (7BRO), **f** 2d ligand–protein interaction of Palbociclib and 3C like main protease (7BRO)

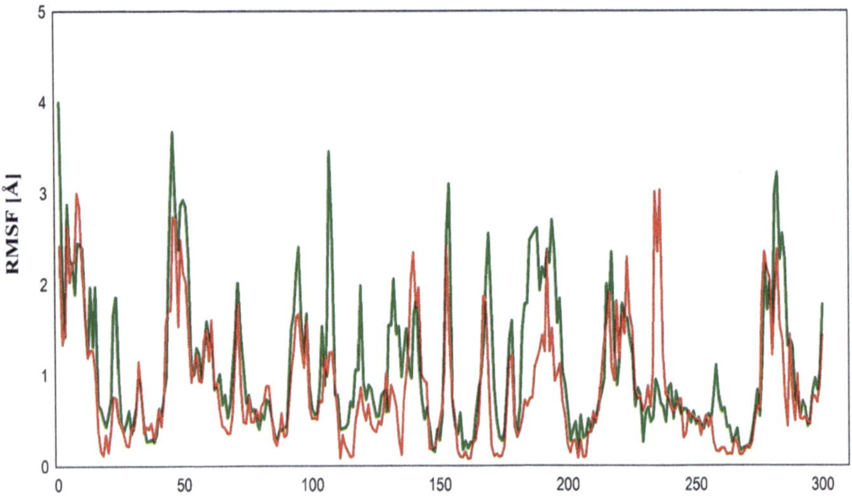

Fig. 6 Molecular dynamics simulations showing RMSF plot of Darunavir (green line) and 3C main protease protein without ligand of SARS-COV-2 (red line)

References

Abdel-Basset, M., Chang, V., & Mohamed, R. (2020). HSMA_WOA: A hybrid novel Slime mould algorithm with whale optimization algorithm for tackling the image segmentation problem of chest X-ray images. *Applied Soft Computing, 95*, 106642.

Abdel-Basset, M., Chang, V., & Nabeeh, N. A. (2021a). An intelligent framework using disruptive technologies for COVID-19 analysis. *Technological Forecasting and Social Change, 163*, 120431.

Abdel-Basset, M., Chang, V., Hawash, H., Chakrabortty, R. K., & Ryan, M. (2021b). FSS-2019-nCov: A deep learning architecture for semi-supervised few-shot segmentation of COVID-19 infection. *Knowledge-Based Systems, 212*, 106647.

Amin, M., & Abbas, G. (2020). Docking study of Chloroquine and Hydroxychloroquine interaction with SARS-CoV-2 spike glycoprotein-An in-silico insight into the comparative efficacy of repurposing antiviral drugs. *Journal of biomolecular structure & dynamics*, 1–11.

Arnott, J. A., Kumar, R., & Planey, S. L. (2013). Lipophilicity indices for drug development. *Journal of Applied Biopharmacy Pharmacokinet, 1*(1), 31–36.

Bapatu, H. R., Maram, R. K., & Murthy, R. S. (2015). Robust and rugged stability-indicating HPLC method for the determination of plerixafor and its related impurities in drug substances. *Journal of Chromatographic Science, 53*(9), 1432–1442.

Braga, S. F., de Melo, L. C., & Barone, P. M. V. B. (2004). Semiempirical study on the electronic structure of antitumor drugs ellipticines, olivacines and isoellipticines. *Journal of Molecular Structure: THEOCHEM, 710*(1–3), 51–59.

Capetti, A. F., Cossu, M. V., Orofino, G., Sterrantino, G., Cenderello, G., Giuseppe, V., & Rizzardini, G. (2017). A dual regimen of ritonavir/darunavir plus dolutegravir for rescue or simplification of rescue therapy: 48 weeks' observational data. *BMC Infectious Diseases, 17*(1), 1–7.

de Oliveira, O. V., Rocha, G. B., Paluch, A. S., & Costa, L. T. (2021). Repurposing approved drugs as inhibitors of SARS-CoV-2 S-protein from molecular modeling and virtual screening. *Journal of Biomolecular Structure and Dynamics, 39*(11), 3924–3933.

Design, L. (2014). Pharmacophore and ligand-based design with Biovia Discovery Studio®.

Dhiman, G., Chang, V., Kant Singh, K., & Shankar, A. (2021). Adopt: automatic deep learning and optimization-based approach for detection of novel coronavirus covid-19 disease using x-ray images. *Journal of Biomolecular Structure and Dynamics*, 1–13.

Elfiky, A. A. (2020). Ribavirin, Remdesivir, Sofosbuvir, Galidesivir, and Tenofovir against SARS-CoV-2 RNA dependent RNA polymerase (RdRp): A molecular docking study. *Life Sciences, 253*, 117592.

Finn, R. S., Martin, M., Rugo, H. S., Jones, S., Im, S. A., Gelmon, K., & Slamon, D. J. (2016). Palbociclib and letrozole in advanced breast cancer. *New England Journal of Medicine, 375*(20), 1925–1936.

Hosseini, F. S., & Amanlou, M. (2020). Simeprevir, potential candidate to repurpose for coronavirus infection: Virtual screening and molecular docking study.

Kasela, S., Daniloski, Z., Bollepalli, S., Jordan, T. X., tenOever, B. R., Sanjana, N. E., & Lappalainen, T. (2021). Integrative approach identifies SLC6A20 and CXCR6 as putative causal genes for the COVID-19 GWAS signal in the 3p21. 31 locus. *Genome Biology*, *22*(1), 1–10.

Kim, D., Lee, J. Y., Yang, J. S., Kim, J. W., Kim, V. N., & Chang, H. (2020). The architecture of SARS-CoV-2 transcriptome. *Cell, 181*(4), 914–921.

Kim, S., Thiessen, P. A., Bolton, E. E., Chen, J., Fu, G., Gindulyte, A., & Bryant, S. H. (2016). PubChem substance and compound databases. *Nucleic Acids Research, 44*(D1), D1202–D1213.

Kouranov, A., Xie, L., de la Cruz, J., Chen, L., Westbrook, J., Bourne, P. E., & Berman, H. M. (2006). The RCSB PDB information portal for structural genomics. *Nucleic Acids Research, 34*(suppl_1), D302-D305.

Kumar, A., Choudhir, G., Shukla, S. K., Sharma, M., Tyagi, P., Bhushan, A., & Rathore, M. (2020). Identification of phytochemical inhibitors against main protease of COVID-19 using molecular modeling approaches. *Journal of Biomolecular Structure and Dynamics*, 1–11.

Kuriata, A., Gierut, A. M., Oleniecki, T., Ciemny, M. P., Kolinski, A., Kurcinski, M., & Kmiecik, S. (2018). CABS-flex 2.0: A web server for fast simulations of flexibility of protein structures. *Nucleic Acids Research*, *46*(W1), W338-W343.

Li, G., Wang, J., He, X., Zhang, L., Ran, Q., Xiong, A., & Zhu, D. (2020a). An integrative analysis identifying transcriptional features and key genes involved in COVID-19. *Epigenomics, 12*(22), 1969–1981.

Li, J., Guo, M., Tian, X., Liu, C., Wang, X., Yang, X., & Liang, Q. (2020b). Virus-host interactome and proteomic survey of PMBCs from COVID-19 patients reveal potential virulence factors influencing SARS-CoV-2 pathogenesis. *BioRxiv*.

Lipinski, C. A., Lombardo, F., Dominy, B. W., & Feeney, P. J. (1997). Experimental and computational approaches to estimate solubility and permeability in drug discovery and development settings. *Advanced Drug Delivery Reviews, 23*(1–3), 3–25.

Lobo-Galo, N., Terrazas-López, M., Martínez-Martínez, A., & Díaz-Sánchez, Á. G. (2021). FDA-approved thiol-reacting drugs that potentially bind into the SARS-CoV-2 main protease, essential for viral replication. *Journal of Biomolecular Structure and Dynamics*, *39*(9), 3419–3427.

Lu, R., Zhao, X., Li, J., Niu, P., Yang, B., Wu, H., & Tan, W. (2020). Genomic characterisation and epidemiology of 2019 novel coronavirus: Implications for virus origins and receptor binding. *The Lancet, 395*(10224), 565–574.

Mitra, D., Pal, A. K., & Mohapatra, P. K. D. (2020). In-silico study of SARS-CoV-2 and SARS with special reference to intra-protein interactions, A plausible explanation for stability, divergency and severity of SARS-CoV-2.

Mitra, D., & Bose, A. (2021). Remarkable effect of natural compounds that have therapeutic effect to stop COVID-19. *RecentAdvances In PharmaceuticalSciences*, 115–126.

Mitra, D., & Mohapatra, P. K. D. (2021a). Cold adaptation strategy of psychrophilic bacteria: an in-silico analysis of isocitrate dehydrogenase. *Systems Microbiology and Biomanufacturing*, 1–11.

Mitra, D., & Mohapatra, P. K. D. (2021b). Discovery of novel cyclic salt bridge in thermophilic bacterial protease and study of its sequence and structure. *Applied Biochemistry and Biotechnology, 193*(6), 1688–1700.

Mitra, D., & Mohapatra, P. K. D. (2021c). Inhibition of SARS-CoV-2 protein by bioactive compounds of edible mushroom; A bioinformatics insight. *International Journal of Advances in Science, Engineering and Technology, 9*(2), 84–88.

Mitra, D., Dey, A., Biswas, I., & Das Mohapatra, P. K. (2021a). Bioactive compounds as a potential inhibitor of colorectal cancer; an insilico study of Gallic acid and Pyrogallol. *Annals of Colorectal Research, 9*(1), 32–39.

Mitra, D., Paul, M., Thatoi, H., & Mohapatra, P. K. D. (2021b). Study of potentiality of dexamethasone and its derivatives against Covid-19. *Journal of Biomolecular Structure and Dynamics*, 1–11.

Muralidharan, N., Sakthivel, R., Velmurugan, D., & Gromiha, M. M. (2021). Computational studies of drug repurposing and synergism of lopinavir, oseltamivir and ritonavir binding with SARS-CoV-2 protease against COVID-19. *Journal of Biomolecular Structure and Dynamics, 39*(7), 2673–2678.

Paraskevis, D., Kostaki, E. G., Magiorkinis, G., Panayiotakopoulos, G., Sourvinos, G., & Tsiodras, S. (2020). Full-genome evolutionary analysis of the novel corona virus (2019-nCoV) rejects the hypothesis of emergence as a result of a recent recombination event. *Infection, Genetics and Evolution, 79*, 104212.

Pathak, G. A., Singh, K., Miller-Fleming, T. W., Wendt, F. R., Ehsan, N., Hou, K., & Mancuso, N. (2021). Integrative genomic analyses identify susceptibility genes underlying COVID-19 hospitalization. *Nature Communications, 12*(1), 1–11.

Pedretti, A., Mazzolari, A., Gervasoni, S., Fumagalli, L., & Vistoli, G. (2021). The VEGA suite of programs: A versatile platform for cheminformatics and drug design projects. *Bioinformatics, 37*(8), 1174–1175.

Rizvi, S. M. D., Shakil, S., & Haneef, M. (2013). A simple click by click protocol to perform docking: AutoDock 4.2 made easy for non-bioinformaticians. *EXCLI Journal, 12*, 831.

Suzuki, T., Otake, Y., Uchimoto, S., Hasebe, A., & Goto, Y. (2020). Genomic characterization and phylogenetic classification of bovine coronaviruses through whole genome sequence analysis. *Viruses, 12*(2), 183.

Tai, Y., Gao, B., Li, Q., Yu, Z., Zhu, C., & Chang, V. (2021). Trustworthy and intelligent covid-19 diagnostic iomt through xr and deep learning-based clinic data access. *IEEE Internet of Things Journal*.

Tilocca, B., Britti, D., Urbani, A., & Roncada, P. (2020). Computational immune proteomics approach to target COVID-19. *Journal of Proteome Research, 19*(11), 4233–4241.

Wang, F., Huang, S., Gao, R., Zhou, Y., Lai, C., Li, Z., & Liu, L. (2020). Initial whole-genome sequencing and analysis of the host genetic contribution to COVID-19 severity and susceptibility. *Cell Discovery, 6*(1), 1–16.

Zhang, M. Q., & Wilkinson, B. (2007). Drug discovery beyond the 'rule-of-five.' *Current Opinion in Biotechnology, 18*(6), 478–488.

Biomedical Data Driven COVID-19 Prediction Using Machine Learning Approach

Arijit Ghosal, Nikita Gupta, Esha Nandi, and Harshita Somolu

Abstract The novel Corona Virus (SARS-CoV-2) is a new strain of Corona Virus that had never affected humans before. The SARS-CoV-2 virus causes the sickness termed COVID-19. COVID-19 has stretched throughout this planet, causing a global pandemic. In this challenging situation, Machine Learning can play an important role in preventing and control COVID-19 by identifying a case, predicting future spreads, and more. It can also play a pivotal role in developing drugs and vaccines to fight the disease. In this work, a machine learning approach has been used to analyze biomedical data of COVID-19 suspected patients to predict whether those patients have positive prediction results or not. To date, there is no proper treatment of COVID-19. Even there is no proper full-proof vaccine also. But this Corona Virus (SARS-CoV-2) can spread from one human to another human without their knowledge through droplets. Minimization of close human interaction is the only possible solution to restrict the spread of the Corona Virus (SARS-CoV-2). Since December 2019, this virus has been spreading throughout the world, involving almost all the countries of the world. In this situation, the machine learning approach can save us by restricting the spread of the virus. Currently there several techniques of tests (like RT-PCR) are there to find whether a person is COVID-19 positive or not. In those tests, nasal and throat swab is collected and then the test is conducted on that. It is costly as well time taking. Moreover, there is a chance that the collector person also may get infected. An alternative machine learning-based approach is proposed in this effort, which does not require close interaction with COVID-19 suspected patients. This approach relies on some symptoms of the suspected patients. As the currently running testing techniques takes a long time to generate a result, there is a possibility that the suspected patient may spread this virus among other persons before the test result come. The proposed scheme minimizes this possibility by providing an early notification of infection before the actual test result comes. If a suspected patient is notified of a positive COVID-19 infection, all precautionary measures like quarantine and contact tracing can be started before the actual test result arrives. This minimizes

A. Ghosal (✉) · N. Gupta · E. Nandi · H. Somolu
Department of Information Technology, St. Thomas' College of Engineering and Technology, Kolkata, West Bengal 700 023, India
e-mail: ghosal.arijit@yahoo.com

V. Chang et al. (eds.), *Artificial Intelligence and Machine Learning Methods in COVID-19 and Related Health Diseases*, Studies in Computational Intelligence 1023,
https://doi.org/10.1007/978-3-031-04597-4_6

the spread of the Corona Virus (SARS-CoV-2) among persons, which reduces the chances of community transmission of this virus. Some common COVID-19 symptoms like—breathing problem, fever, dry cough, sore throat, headache, fatigue is considered in this effort. Some other factors which increase the possibility of being infected by COVID-19 like abroad travel history, close contacts with COVID-19 positive patients, attending large public gatherings, visiting public exposed places, whether a family person working in public exposed places, wearing a mask, sanitization after coming from the market are also considered in this effort along with information of COVID-19 symptoms. A total of 15 factors are considered in this work. This work aims to provide support to society by predicting the possibility of COVID-19 positive before the actual test result comes without any close interaction with the suspected person.

Keywords COVID-19 · Machine learning technique · Symptoms · Prediction

1 Introduction

A pandemic that has currently taken over the whole world is caused by the newly discovered Corona Virus (COVID-19). The origin and source of the SARS-CoV-2 remain unknown. But a suspected bat origin was suggested after 96% genome sequence identity was demonstrated between SARS-CoV-2 and another Corona Virus named Bat-CoV-RaTG13 isolated from bat species that colonized a province nearly 2000 km away from Wuhan.

The most common warning signs of COVID-19 are Fever, Dry cough and fatigue. Other symptoms that are less common and may affect some patients include Loss of taste or smell, Nasal congestion as well as vomiting, Conjunctivitis or red eyes, Sore throat, Headache, Chills or dizziness, etc. Populace aged above 60 years, and populace who are already suffering from different chronic diseases like heart and lung problems, high blood pressure, obesity, diabetes, or malignancy (cancer), are at high threat of serious illness. Any person having the symptoms mentioned above should take up the COVID-19 test. It is advised by the doctors because even if there are no symptoms, but they can have the disease as they have been in close contact with the patient and for an airborne disease like this, it is very natural to be transmitted.

It is very important to know what kind of tests should be done to beware of the disease. There are two kinds of tests like the VIRAL TEST and ANTIBODY TEST. It is also known that in the majority state of affairs, a molecular investigation is exploited to become aware of SARS-CoV-2 and corroborate infection. Polymerase Chain Reaction or PCR is the universally exploited molecular investigation. There is something called the Rapid antigen tests or RDT, which detect viral proteins known as antigens. The tests are cheaper than PCR and will offer more quickly, but they are not accurate. The measures that doctors suggest includes cleaning hands often, coughing or sneezing in bending the elbow—not hands, avoid touching eyes, nose and mouth, limiting social gatherings and time spent in crowded places, avoiding close contact

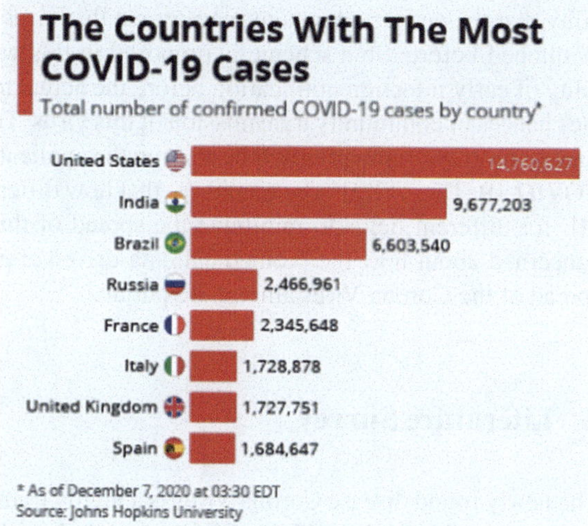

Fig. 1 The countries with most COVID-19 cases [7 December 2020]

with someone who is sick, cleaning and disinfecting frequently touched objects and surfaces. The measures are all important, but explaining the situation in one word is being in quarantine or isolation.

Globally in the past week, cases of COVID-19 have remained at a large amount, while new deaths have continued to be at an alarming level. This conveys the cumulative numbers to above 66 million informed infections and 2 million demises internationally since the pandemic outbreak. As of now, the whole world has not yet recovered and total cases have been 69,038,276, deaths have been 1,571,803 and recovered has been 47,835,263. The active cases are still 19,631,210 (Lalmuanawma et al., 2020).

The top ten countries on the point of view of the entire count of COVID-19 infections by different countries worldwide until 7 December 2020 are informed through the above mentioned Fig. 1. There is an urge for inventive clarifications to build up, cope as well as analyze colossal records on the growing set of connections of subjects with contagion, patient minutiae, community progress of those, and amalgamate with therapeutic tryouts and, pharmaceutical, genomic furthermore civic health information (Ong et al., 2020). Through machine learning techniques, mock-ups are put up exploiting chronological statistics and these mock-ups are exercised to forecast the novel upshot. Classification, regression, grouping, deep learning, etc. are a few of the machine learning techniques utilized in assorted application areas, for instance, analysis of the image, health informatics, speech recognition, etc. Assorted information sets concerning this have been opened and made public. The no-contact policy is catalyzed by the use of technology and web applications where text, diagrams, and lengthy reports can be created, uploaded, and analyzed. There are many machine learning model implications for developing the execution of machine learning schemes for the prophecy of authenticated incidents of COVID-19

(Rao et al., 2020). As tests take a long time to generate a result, there is a possibility that the suspected patient may spread this virus among other persons, as it is mentioned before. So a scheme is proposed in this work that minimizes the possibility of early infection notification before the actual test result comes. This reduces the chances of community transmission of this virus. The next section is the literature survey; few research works will be seen on the application of ML methods to prevent COVID-19. This will also be seen there that how different approaches are taken with ML for different fields to minimize the spread of this virus. This paper is mainly concerned about how the biomedical data-driven approach will help minimize the spread of the Corona Virus among the public.

2 Literature Survey

The newly found disease Corona Virus is a highly transmittable and unhealthy infectious virus with no quantifiable agreed antiviral medicine or vaccine available for the healing. Machine Learning (ML) can be used to help the early-stage prediction of COVID-19 cases and reduce its spread.

Using an AI approach to predict the epidemic, a study was done using a modified Susceptible-Exposed-Infectious-Removed (SEIR) model. The results showed a peak of affected people. By using this prediction, safety measures can be implemented (Yang et al., 2020). Due to the need for indispensable data and indecision, epidemiological mock-ups are confronted concerning the release of superior correctness for enduring forecasts. Being an option to the Susceptible-Infected-Resistant or SIR depended on mock-ups, this learning suggests an amalgam ML approach (Pinter et al., 2020). The further superior adaptation of SIR-based mock-ups cautiously judges the crucial dynamics in addition to the steady populace (Miller, 2017). For case in point, at the incidence of enduring imperviousness conjecture when the imperviousness is not realized, leading revival from contagion, the Susceptible-Infectious-Susceptible (SIS) mock-up was advocated (Werkman et al., 2011). Allen Institute for AI collaborated with top investigating associations and released an open basis, weekly revised data set. The dataset included epidemiological, clinical, and genetic data analyzed using AI and updated to help diagnose and prevent Corona Virus (Alimadadi et al., 2020).

The clinical chemistry team and Laboratory Medicine have proposed a method using ML mock-ups for COVID-19 revealing, anchored in regular blood examinations. Machine learning method can be enforced on blood examinations as an appendage and unconventional way for rRT-PCR to quickly and gainfully recognize COVID-19 (Cabitza et al., 2021). ML and soft computing models such as multi-layered perceptron or MLP and adaptive association-based fuzzy implication coordination, ANFIS, are used. An effective prophecy model is created using these (Ardabili et al., 2020). Although RT-PCR stays the orientation measure for formulating a definitive diagnosis for COVID-19 contagion (Wong et al., 2020), huge false-negative amount (Chan et al., 2020), the lack of accessibility of the test in the premature phase of the epidemic confined timely finding of patients contagious. Chest CT can help identify the early phase of lung infection and prompt larger public health surveillance and response systems (Zu et al., 2020).

Studies made all through a pandemic are intrinsically fettered by synthetically huge sickness pervasiveness and the opted character of contributors, whose syndrome sternness permitted hospice admittance in addition to CT assessment. Idyllically, algorithms are necessitated to be coached on the complete spectrum of illness, together with asymptomatic as well as early-phase incidents, if CT elucidation by machine learning technique can be enforced to factual-world statistics (Bachtiger et al., 2020). Skilled radiologists detected that COVID-19 exhibits different behavior than erstwhile viral pneumonia from CT images. ML schemes should be put into practice on CT images of the abdomen, X-ray images of the chest, blood examination upshots on a proper dataset to get the optimum results. The Internet plays a massive role in everyone's lives (Barstugan et al., 2020).

A deep learning-based facet mining framework for involuntary COVID-19 detection can also be used. DenseNet, MobileNet, ResNet, Xception, InceptionResNetV2, InceptionV3, etc. were chosen amid a puddle of deep convolutional neural networks to feed several machines and to categorize issues as either a case of COVID-19 positive or not (Kassania et al., 2021). A combination of the mechanistic and machine-learning methods is presented. Improved ARGONet exploits a data-controlled scheme to integrate efforts from official healthiness testimonies (Liu et al., 2020). by using a sample dataset, an extrapolative prophecy approach can be built depending on the XGBoost machine learning procedure, envisaging the death menace and portraying a quantifiable road to detecting crucial incidents from several incidents (Yan et al., 2020). Characteristics of patients such as age, co-morbidities and varied clinical symptoms can help classify the level of infection severity; predict the disease outcome and the need for hospitalization. ML techniques fed with demographic and clinical data can assist in prognostic prediction and physician decision-making (De Souza et al., 2021). A team of IEEE presents four standard forecasting models: Linear Regression or LR, Least Absolute Shrinkage and Selection Operator or LASSO, Exponential Smoothing or ES, and Support Vector Machine or SVM (Rustam et al., 2020).

Ahmad et al. (2021) proposed the following taxonomy of ML methodologies used for prediction: habitual deep learning regression, ML regression, investigation of the network, search queries-based method. Assaf et al. (2020) suggested different ML mock-ups such as artificial neural networks, Random Forest-based categorization, Classification and Regression Tree, or CRT decision tree model. The prime upshot was a threat for complicated illness, labeled as mechanical aeration, multi-organ collapse, admittance to the ICU, and/or death. The American Heart Association suggests that among the patients with COVID-19, there is a high prevalence of cardiovascular disease and they are at a higher risk of morbidity and mortality. Continuation of clinically designated angiotensin-adapting enzyme inhibitors and angiotensin receptor obstructer prescriptions is recommended depending on the existing proof at this time (Clerkin et al., 2020). Apart from detecting patients, it is also important to distinguish between who is affected more and who is affected less. Tárnok

(2020) suggests the use of ML for single cell data. In truth, effortless to apply and quick examines drawn from Cytomics or multi-OMICS techniques are required to discriminate between severe and mild cases quickly.

Rao et al. (2020) proposed a mobile-based web survey in cities using ML. The collection of data from an individual will include location details, demographic information, voyage record in the preceding 14 days, and record of contact with a COVID-19 patient in the last 14 days, symptoms and signs if sick. Northwell COVID-19 Research Consortium suggests evidence-based prediction tools help clinical decision-making, making medical help faster and easier (Debnath et al., 2020). Tim Mackey et al. proposed that by use of ML and big data. The aim of this learning was to identify and typify user-engendered discussions that might be correlated with COVID-19 by collecting data from Twitter accounts across the world (Mackey et al., 2020). This will help people to take care of their fear amidst this pandemic. There was an analysis done about what people are discussing; their concerns and sentiments can be done using the ML approach and Latent Dirichlet Allocation (LDA). This real-time monitoring helps analyze the pandemic situation among people (Xue et al., 2020). A different approach suggests a ML model uncover hidden patterns based on reported cases. This model depended on facet lessening to discover factors that were major elements for baring secret models and prophetic investigation utilized an unprejudiced hierarchical approach exercising the Bayesian estimator scheme to conjecture precedent infectivity from existing casualties (Vaid et al., 2020). Classification of the pathogens can help in the identification of this disease. A configuration-liberated scheme depended on fundamental genomic identities using ML that is able to convey decidedly perfect data to identify pathogens and thus identify patients affected by COVID-19 (Randhawa et al., 2020).

Mele and Magazzino et al. (2021) has pointed out a relationship between pollution emissions, economic growth and deaths due to COVID-19. By using a dataset, variables are expanded and casualty analysis is done using ML. An approach of Corona Virus vaccine design using reverse vaccinology and machine learning is presented. The two primary design strategies are: the practice of the entire virus or hereditarily brought out vaccine antigens that might be conveyed employing diverse arrangements (Ong et al., 2020). Then comes another approach that takes eight different studies: The first study applies an improved mathematical model to analyze and predict the growth of the epidemic. They have shown here iterative weighting, which is used for correcting Generalized Inverse Weibull circulation. This was orchestrated for the cloud computing podium for the actual prophecy of the growing behavior of this pandemic (Tuli et al., 2020). Secondly, another study proposed a prejudiced derivative regression and nonlinear machine learning (PDR–NML) technique for the global pandemic prophecy of COVID-19. Two dissimilar representations were utilized explicitly, PPDLR for standardizing the facets for appropriate prophecy and PDLFR for accurate predictions (Kavadi et al., 2020). A novel Support Vector Regression method is used to analyze five different tasks related to the novel Corona Virus. In this exertion, machine learning is used. Rather than effortless regression stripe, they also employ support vectors to acquire enhanced categorization exactness (Yadav et al., 2020). The experimental study shows that weather variables are more pertinent in predicting the death rate than the former census changeable, such as age, population, urbanization, etc. (Malki et al., 2020).

Next, learning aims to contribute to the literature by performing clustering analyses employing k-means and hierarchic clustering schemes. It showed that the optimum number of clusters for 142 countries is three (Aydin & Yurdakul, 2020). In addition, a hybrid model using deep and classical machine learning for face mask detection is presented. The advised representation consists of two ingredients: primary is to the extraction of the trait using Resnet50, another is for the categorization procedure of face masks employing Support Vector Machine or SVM, ensemble algorithms and decision trees (Loey et al., 2021). In the next study, a ML-based time series prediction model is used to derive the epidemic curve and predict the trend of the epidemic. As per the mathematical assessment, the global outburst of the pandemic will be in zenith in delayed October, with an approximation of 14.12 million populace infected cumulatively (Wang et al., 2020). The research activity also discusses the suggestions that convey people on mock-up plans, medicinal connoisseurs and policy-creators in the present circumstances while dealing with the COVID-19 pandemic (Lalmuanawma et al., 2020). Weekly epidemiological update of WHO dated 8 December 2020 is available at (Weekly epidemiological update-8 December 2020, 2021). A data set (Symptoms and COVID Presence (May 2020 data), 2021) is storing the biomedical information collected from the report released by WHO during June 2020. COVID-19 Radiography Database (COVID-19 Radiography Database, 2021) consists of X-Ray pictures of 1143 COVID-19 confirmed cases, 1341 regular images in addition to 1345 viral pneumonia pictures. This database is used in this work for comparative analysis. This database is the winner of the COVID-19 Dataset Award by Kaggle Community. Srivastava (Srivastava, 2014) has recommended Weka Tool for different purposes, including classification. X-Ray image-based COVID-19 identification scheme (Dhiman et al., 2021) has also been proposed. A deep learning-based approach has been adopted in this case. Convolutional neural network-based (CNN) models have been used in this effort. They have also used Decision Tree (J48) for the identification of COVID-19 infected persons. In the Extended Reality (XR) and Deep Learning-based approach (Tai et al., 2021), clinic data have been accessed. ACGAN-depended smart forecasting algorithm has been used there for training the system.

3 Proposed Methodology

SARS-CoV-2, which is universally acknowledged as Corona Virus, is a new strain of Corona Virus. As it is a new strain in the world, so we do not have any prior knowledge about this virus. We do not have knowledge of any characteristics of this virus also so that we can say anything about its characteristics. This virus is causing the disease named COVID-19. This disease is spreading from one human being to other humans rapidly. This way, it has created a global pandemic across the globe and created a health emergency situation.

Identification of populace diagnosed with SARS-CoV-2 is the main decisive job to curtail the opportunity of populace getting contaminated of the virus. Every infectious disease has some specific symptoms which help to identify that disease. In the case of COVID-19, identification of symptoms was very tough at the initial phase as we did not have any prior knowledge about the disease as it was the first time in the world. After almost one year of its first infection, we have developed some knowledge about the symptoms of the disease. World Health Organization (WHO), Centers for Disease Control and Prevention (CDC) have also guided us about the different symptoms of COVID-19 from time to time. Prediction of infection of COVID-19, that is, whether a person is infected with Corona Virus (SARS-CoV-2) or not, is very much crucial and essential as this will help us stop further spreading of that virus. Eventually, that will also help us to control the spread of COVID-19 by breaking the contamination chain. It is known that machine learning can be applied where there is no clear rule of any decision-making. In the case of identification or prediction of COVID-19 infected people, it is very tough to form a clear rule. Hence, machine learning can play an important role in identifying or predicting SARS-CoV-2 infected people as it does not require any close interaction with the suspected Corona Virus (SARS-CoV-2) infected people.

Lots of researches have already been explored by different researchers aiming to predict COVID-19 through machine learning by applying different features. Some researchers are trying to predict COVID-19 based on cough sound. But due to the lack of proper data set, this path of exploration is still in the almost scratch situation. IISc Bengaluru of India is trying to identify whether a person is contagious with COVID-19 or not employing cough sound. Many researchers have used facets computed from medical images like X-Ray figures etc., to predict COVID-19. Prediction of COVID-19 should take less time because time is a big factor in preventing the transmission of the sickness. The late prediction will increase the possibility of spreading the disease more among the people. Prophecy of COVID-19 using medical images like X-ray images will take a long time. This will increase the risk of spreading the virus. Moreover, there is a high risk of getting infected by the Corona Virus (SARS-CoV-2) in the case of the health workers while dealing with suspected patients for X-Ray images and purposes. The close vicinity of suspected patients should be avoided in this case as much as possible. So medical image-based approach increases the risk of getting infected by the virus for health workers and increases the spreading of Corona Virus or SARS-CoV-2 among common people as this is a time-consuming process. Currently, COVID-19 is being identified through mostly RT-PCR tests. This test takes a long time—almost a day, to produce the final result. But there is no assurance that the final result is fully correct. So the suspected person may need to undergo this test multiple times indifference case. So there is a need to find a solution that will predict whether a suspected person is COVID-19 infected or not.

The proposed scheme relies on a machine learning approach using the biomedical information and past activities of the suspected people. This biomedical information contains information on different symptoms and co-morbidity factors of suspected

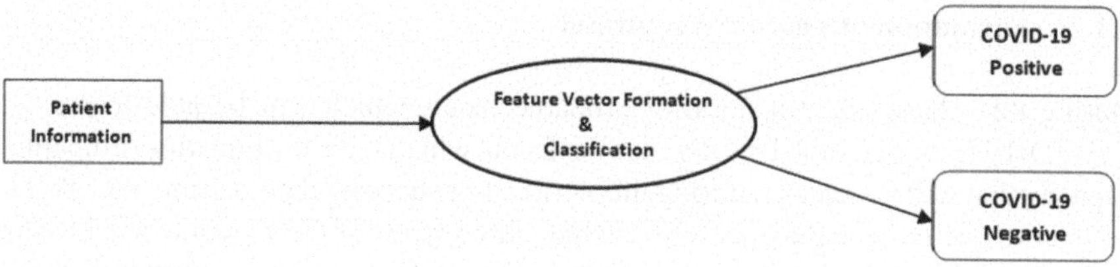

Fig. 2 Flow chart of COVID-19 prediction using machine learning approach

patients. As per WHO, certain guideline activities which increase the possibility of being infected by COVID-19 within the last 14 days are considered in this work. This symptom, past activities and co-morbidity factors-based approach are much quicker and much better (in terms of accuracy) than the medical image-based approach for prediction of COVID-19. The proposed approach is based on symptoms, past activities, and co-morbidity factors of the suspected person through comparative analysis. Also, this approach does not require close vicinity of the suspected patient, which significantly reduces the possibility of being infected by COVID-19. The proposed scheme is reflected through the above Fig. 2.

3.1 Feature Computation

Computation of facet is the main important task of any machine learning-based scheme. This step can be treated as the heart of any machine learning-based scheme because this step writes the path of performance of the system. It is true for any machine learning-based scheme that better will be the facet selection and computation will be better the discrimination accuracy of the system. Better discrimination accuracy means better performance of the system. If proper facets are selected and computed, then the task of any classifier gets simple. Any classifier will be able to show brilliant performance because of the excellent facet set. So the facets need to be chosen judiciously. There is an acute shortage of benchmark COVID-19 data set as this virus is totally new to us. In this exertion, the recordset is collected from the information released by WHO during June 2020 through the website of kaggle.com (Symptoms and COVID Presence (May 2020 data), 2021a). The best feature set for COVID-19 prediction has been suggested from the data set through experimental studies. All the features mentioned in the data set consist of medical symptoms experienced by the suspected patients and their different past activities. WHO and CDC have prescribed certain past activities which increase the risk of getting infected by the Corona Virus (SARS-CoV-2). These past activities are also considered in the facet setting.

3.1.1 Formation of Feature Vector/Set

During this phase, identification of probable facets which will be able to predict COVID-19 is carried out. This denotes the facets which have the high discriminating capacity that will be selected during this period. For successful prediction of COVID-19 feature vector or feature set must include three types of facets while considering biomedical data-driven facets. These three types of facets are—symptoms-based facets, co-morbidity-based features and facets based on past activities of the last 14 days. All these three types of facets are considered in this work and these three categories of facets play a vital role in the prediction of COVID-19.

The dataset, which is collected from the report released by WHO during June 2020 through the website of kaggle.com (Weekly epidemiological update-8 December 2020, 2021b), consists of 20 facets, including all the said three types of facets—symptoms based facets, co-morbidity based features and facets based on past activities of last 14 days. There are seven symptoms based facets (breathing, fever, dry cough, sore throat, running nose, headache and fatigue), six co-morbidity based features (asthma, chronic lung disease, heart disease, diabetes, hypertension and gastrointestinal) and seven facets which are formed based on past activities of last 14 days (abroad travel, contact with COVID-19 patient, attendance in a large gathering, visiting public exposed places, family member working in public exposed places, wearing masks and sanitization after from market). This feature set is denoted as F_1, which is 20 dimensional.

At the end of this phase, the final feature set gets delineated after applying the proper facet assortment procedure. Several techniques exist for selecting facets like—Sequential Backward Selection or SBS, Sequential Forward Selection or SFS, Plus l-take away r Selection, Sequential Floating Backward Search or SFBS, Sequential Floating Forward Search or SFFS and Max–Min approach in the domain of machine learning. In this work, Sequential Floating Backward Search (SFBS) has been considered for the final formation feature set by selecting the best possible set of facets. Sequential Floating Backward Search or SFBS approach begins with all possible assemblies of facets and then less irrelevant facets are dropped to form the final feature set. According to WHO, breathing, fever, dry cough are the dreadfully widespread warning signs of COVID-19. The other symptoms: sore throat, running nose, headache, and fatigue, are considered less common symptoms. Applying the concept of Sequential Floating Backward Search (SFBS), a less significant feature (headache) is dropped in the next step to form the best optimal facet set. Similarly, asthma being a less significant co-morbidity feature, along with abroad travel being a less significant facet based on past activities of the last 14 days as per WHO, is dropped in the next step of SFBS. Hence the reduced dimensional feature set consists of 6 symptoms based facets (breathing, fever, dry cough, sore throat, running nose and fatigue), five co-morbidity based features (chronic lung disease, heart disease, diabetes, hypertension and gastrointestinal), and 6 facets which are formed based on past activities of last 14 days (contact with COVID-19 patient, attendance in a large gathering, visiting public exposed places, family member working in public exposed places, wearing masks and sanitization after from market). This feature set is denoted as F_2, which is 17 dimensional. As per the SFBS approach of feature selection, there is always a scope of return of a facet in the previous facet assembly if categorization accuracy drops down after removing the feature.

Performance of these two features (F_1 and F_2) is carried out using Naïve Bayes, Multi-Layer Perceptron (MLP) model and Support Vector Machine (SVM). Analysis of performance is discussed during the experimental results section.

3.2 Classification

The classifier is an integral part of any machine learning exertion. They play an important role in machine learning because they do the classification task with the help of the facet set generated during the feature formation phase. The differentiating capabilities of both two features (F_1 and F_2) are assessed through Naïve Bayes, Multi-Layer Perceptron (MLP) model and Support Vector Machine (SVM) classifier. For the implementation of Naïve Bayes, Multi-Layer Perceptron (MLP) model and Support Vector Machine (SVM) classifier weka-tool (COVID-19 Radiography Database, 2021) have been used in this work.

The dataset consists of biomedical information from 5434 people. This dataset is applied both for testing and training roles by dividing the dataset into two identical divisions. Each division will have biomedical information of 2717 people.

MLP (Multi-Layer Perceptron) model of Neural Network has been applied in this work, considering the quantity of neurons in the input stratum identical to the dimension of facet setting. Hence for F_1, the numbers of neurons in the input stratum were 20 (depicted in Fig. 3) and for F_2, the numbers of neurons in the input stratum were 17 (depicted in Fig. 4). The output echelon is always made of two neurons irrespective of the feature set, which indicates the number of output categories. In this work output category is always two—COVID-19 positive or negative.

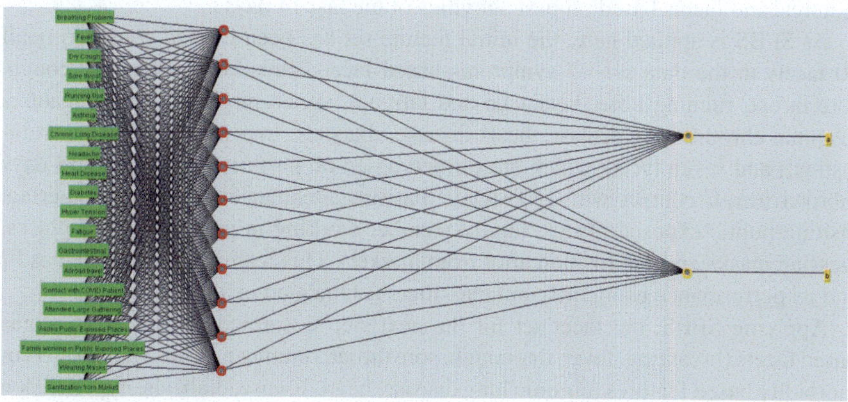

Fig. 3 Configuring MLP for F_1 (a = COVID-19 positive, b = COVID-19 negative)

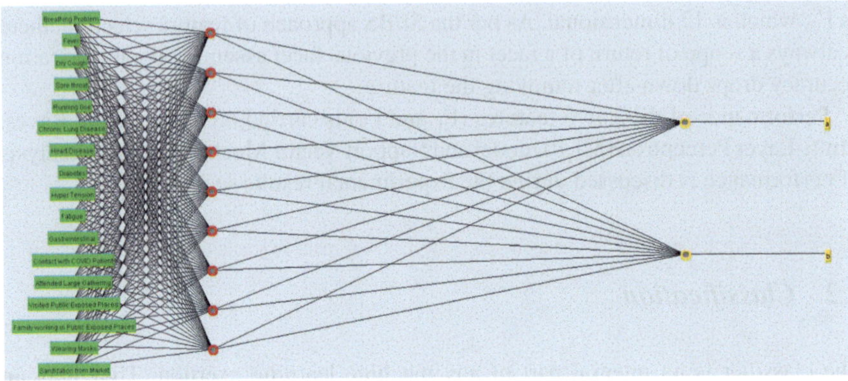

Fig. 4 Configuring MLP for F_2 (a = COVID-19 positive, b = COVID-19 negative)

Tenfold cross-validation is applied in this work to enforce Naïve Bayes as well as Support Vector Machine (SVM). During the implementation of Support Vector Machine (SVM), "Quadratic" kernel is exercised to attain the best classification result.

4 Experimental Results

The whole experiment is carried out with data sets. The dataset is assembled from the report released by WHO in June 2020 through the website of kaggle.com (Symptoms and COVID Presence (May 2020 data), 2021a). The data sets consist of some biomedical information of 5434 people. The Sequential Floating Backward Search (SFBS) technique is applied to uncover the best facet set. The data set consists of three types of biomedical information—symptoms-based facets, co-morbidity-based features, and facets based on past activities of the last 14 days.

As SFBS is applied here, the initial feature set has been formed considering all 20 facets in the data set—7 symptoms based facets (breathing, fever, dry cough, sore throat, running nose, headache and fatigue), six co-morbidity based features (asthma, chronic lung disease, heart disease, diabetes, hypertension and gastrointestinal) and seven facets which are formed based on past activities of last 14 days (abroad travel, contact with COVID-19 patient, attendance in a large gathering, visiting public exposed places, family member working in public exposed places, wearing masks and sanitization after from market). This feature set is denoted as F_1 and its performance using different classifiers is tabulated in Table 1.

Applying SFBS, the facet set for the next step is formed having 6 symptoms based facets (breathing, fever, dry cough, sore throat, running nose and fatigue), 5 co-morbidity based features (chronic lung disease, heart disease, diabetes, hypertension and gastrointestinal) and six facets which are formed based on past activities of last

Table 1 Accuracy of prediction of COVID-19 (using F_1)

Classifier	Discrimination exactness (in %)
Naïve bayes	96.4115
MLP (multi-layer perceptron)	98.0493
SVM (support vector machine)	96.7427

Table 2 Accuracy of prediction of COVID-19 (using F_2)

Classifier	Discrimination exactness (in %)
Naïve Bayes	94.9761
MLP (multi-layer perceptron)	97.8653
SVM (support vector machine)	95.6386

14 days (contact with COVID-19 patient, attendance in a large gathering, visiting public exposed places, family member working in public exposed places, wearing masks and sanitization after from market) after removing headache from symptoms based facets, asthma from co-morbidity based features and abroad travel from facets formed based on past activities of last 14 days. This facet set is denoted as F_2 and its performance using different classifiers is tabulated in Table 2.

From Table 1 as well as Table 2, this is detected that F_1 performs better than F_2. Hence, F_1 is the recommended facet set.

4.1 Comparative Study

A comparative study is also carried out to compare the proposed biomedical information-based facet set with other data sets. A dataset (Symptoms and COVID Presence (May 2020 data), 2021) is used for comparison purposes. This data set is storing X-Ray figures of 1143 COVID-19 confirmed incidents, 1341 ordinary images in addition to 1345 viral pneumonia figures. Deep Learning is applied there to find out the performance of the dataset by applying the deep learning concept.

The comparative analysis is tabulated in Table 3. The relative investigation shows that the advised biomedical information-based facet set performs better than the X-Ray-based facet for predicting COVID-19.

Table 3 Comparative Analysis of Accuracy of Prediction of COVID-19

Classifier	Classification accuracy (in %)
MLP (multi-layer perceptron) in (symptoms and COVID presence (May 2020 data), 2021)	92.9724
Proposed method (MLP) using F_1	98.0493

Also biomedical information based facet set performs quicker than facets computed from X-Ray images. Experimentally, it is observed that the average response time for the X-Ray image-based approach is 3.929 s and for biomedical information based it is 0.128 s only.

5 Conclusion

This work indicates that the suggested biomedical information-based facet set can predict COVID-19 with a 98.0493% success rate using Multi-Layer Perceptron (MLP) classifier. This work also establishes that the biomedical information-based facet set is much quicker than the X-Ray image-based system. In the future, other facets may be explored to predict COVID-19 more accurately. Cough sounds of patients may play an important role in predicting COVID-19, but there is a very much lack of benchmark COVID-19 dataset.

References

Ahmad, A., Garhwal, S., Ray, S. K., Kumar, G., Malebary, S. J., & Barukab, O. M. (2021). The number of confirmed cases of covid-19 by using machine learning: Methods and challenges. *Archives of Computational Methods in Engineering, 28*

Alimadadi, A., Aryal, S., Manandhar, I., Munroe, P.B., Joe, B., & Cheng, X. (2020) Artificial intelligence and machine learning to fight COVID-19.

Ardabili, S. F., Mosavi, A., Ghamisi, P., Ferdinand, F., Varkonyi-Koczy, A. R., Reuter, U., Rabczuk, T., & Atkinson, P. M. (2020). Covid-19 outbreak prediction with machine learning. *Algorithms, 13*(10), 249.

Assaf, D., Gutman, Y.A., Neuman, Y., Segal, G., Amit, S., Gefen-Halevi, S., Shilo, N., Epstein, A., Mor-Cohen, R., Biber, A., & Rahav, G., (2020). Utilization of machine-learning models to accurately predict the risk for critical COVID-19. *Internal and emergency medicine, 15*(8), 1435–1443.

Aydin, N., & Yurdakul, G. (2020). Assessing countries' performances against COVID-19 via WSIDEA and machine learning algorithms. *Applied Soft Computing, 97*, 106792.

Bachtiger, P., Peters, N. S., & Walsh, S. L. (2020). Machine learning for COVID-19—asking the right questions. *The Lancet Digital Health, 2*(8), e391–e392.

Barstugan, M., Ozkaya, U., & Ozturk, S. (2020). Coronavirus (covid-19) classification using ct images by machine learning methods. ArXiv preprint arXiv:2003.09424.

Cabitza, F., Campagner, A., Ferrari, D., Di Resta, C., Ceriotti, D., Sabetta, E., Colombini, A., De Vecchi, E., Banfi, G., Locatelli, M., & Carobene, A. (2021). Development, evaluation, and validation of machine learning models for COVID-19 detection based on routine blood tests. *Clinical Chemistry and Laboratory Medicine (CCLM), 59*(2), 421–431.

Chan, J. F. W., Yuan, S., Kok, K. H., To, K. K. W., Chu, H., Yang, J., Xing, F., Liu, J., Yip, C. C. Y., Poon, R. W. S., & Tsoi, H. W. (2020). A familial cluster of pneumonia associated with the 2019 novel coronavirus indicating person-to-person transmission: A study of a family cluster. *The Lancet, 395*(10223), 514–523.

Clerkin, K. J., Fried, J. A., Raikhelkar, J., Sayer, G., Griffin, J. M., Masoumi, A., Jain, S. S., Burkhoff, D., Kumaraiah, D., Rabbani, L., & Schwartz, A. (2020). COVID-19 and cardiovascular disease. *Circulation, 141*(20), 1648–1655.

De Souza, F.S.H., Hojo-Souza, N.S., Dos Santos, E.B., Da Silva, C.M., & Guidoni, D.L. (2021). Predicting the disease outcome in COVID-19 positive patients through Machine Learning: a retrospective cohort study with Brazilian data. *Frontiers in Artificial Intelligence, 4*.

Debnath, S., Barnaby, D. P., Coppa, K., Makhnevich, A., Kim, E. J., Chatterjee, S., Tóth, V., Levy, T. J., & d Paradis, M., Cohen, S.L. and Hirsch, J.S. (2020). Machine learning to assist clinical decision-making during the COVID-19 pandemic. *Bioelectronic Medicine, 6*(1), 1–8.

Dhiman, G., Chang, V., Kant Singh, K. and Shankar, A., 2021. ADOPT: automatic deep learning and optimization-based approach for detection of novel coronavirus COVID-19 disease using X-ray images. *Journal of Biomolecular Structure and Dynamics*, 1–13.

Kaggle.com. 2021a. Symptoms and COVID Presence (May 2020 data). [online] Available at: <https://www.kaggle.com/hemanthhari/symptoms-and-covid-presence> [Accessed 14 September, 2021a].

Kaggle.com. 2021b. COVID-19 radiography database. [online] Available at: <https://www.kaggle.com/tawsifurrahman/covid19-radiography-database> [Accessed 14 September, 2021b].

Kassania, S. H., Kassanib, P. H., Wesolowskic, M. J., Schneidera, K. A., & Detersa, R. (2021). Automatic detection of coronavirus disease (COVID-19) in X-ray and CT images: A machine learning based approach. *Biocybernetics and Biomedical Engineering, 41*(3), 867–879.

Kavadi, D. P., Patan, R., Ramachandran, M., & Gandomi, A. H. (2020). Partial derivative nonlinear global pandemic machine learning prediction of covid 19. *Chaos, Solitons & Fractals, 139*, 110056.

Lalmuanawma, S., Hussain, J., & Chhakchhuak, L. (2020). Applications of machine learning and artificial intelligence for Covid-19 (SARS-CoV-2) pandemic: A review. *Chaos, Solitons & Fractals, 139*, 110059.

Liu, D., Clemente, L., Poirier, C., Ding, X., Chinazzi, M., Davis, J.T., Vespignani, A., & Santillana, M. (2020). A machine learning methodology for real-time forecasting of the 2019–2020 COVID-19 outbreak using Internet searches, news alerts, and estimates from mechanistic models. ArXiv preprint arXiv:2004.04019.

Loey, M., Manogaran, G., Taha, M. H. N., & Khalifa, N. E. M. (2021). A hybrid deep transfer learning model with machine learning methods for face mask detection in the era of the COVID-19 pandemic. *Measurement, 167*, 108288.

Mackey, T., Purushothaman, V., Li, J., Shah, N., Nali, M., Bardier, C., Liang, B., Cai, M., & Cuomo, R. (2020). Machine learning to detect self-reporting of symptoms, testing access, and recovery associated with COVID-19 on Twitter: Retrospective big data infoveillance study. *JMIR Public Health and Surveillance, 6*(2), e19509.

Malki, Z., Atlam, E. S., Hassanien, A. E., Dagnew, G., Elhosseini, M. A., & Gad, I. (2020). Association between weather data and COVID-19 pandemic predicting mortality rate: Machine learning approaches. *Chaos, Solitons & Fractals, 138*, 110137.

Mele, M., & Magazzino, C. (2021). Pollution, economic growth, and COVID-19 deaths in India: A machine learning evidence. *Environmental Science and Pollution Research, 28*(3), 2669–2677.

Miller, J. C. (2017). Mathematical models of SIR disease spread with combined non-sexual and sexual transmission routes. *Infectious Disease Modelling, 2*(1), 35–55.

Ong, E., Wong, M. U., Huffman, A., & He, Y. (2020). COVID-19 coronavirus vaccine design using reverse vaccinology and machine learning. *Frontiers in Immunology, 11*, 1581.

Pinter, G., Felde, I., Mosavi, A., Ghamisi, P., & Gloaguen, R. (2020). COVID-19 pandemic prediction for Hungary: A hybrid machine learning approach. *Mathematics, 8*(6), 890.

Randhawa, G. S., Soltysiak, M. P., El Roz, H., de Souza, C. P., Hill, K. A., & Kari, L. (2020). Machine learning using intrinsic genomic signatures for rapid classification of novel pathogens: COVID-19 case study. *PLoS ONE, 15*(4), e0232391.

Rao, A.S.S., Miller, D.D., Berman, A.E., Hess, D.C., & Krantz, S.G. (2020). Immediate and near future prediction of COVID-19 patients in the US population aged 65+ with the prior medical conditions of hypertension, cardiovascular and lung diseases: methods, models and acute care estimates. *medRxiv*.

Rustam, F., Reshi, A. A., Mehmood, A., Ullah, S., On, B. W., Aslam, W., & Choi, G. S. (2020). COVID-19 future forecasting using supervised machine learning models. *IEEE Access, 8*, 101489–101499.

Srivastava, S., (2014). Weka: a tool for data preprocessing, classification, ensemble, clustering and association rule mining. *International Journal of Computer Applications 88*(10).

Tai, Y., Gao, B., Li, Q., Yu, Z., Zhu, C. and Chang, V., (2021). Trustworthy and Intelligent COVID-19 Diagnostic IoMT through XR and Deep Learning-based Clinic Data Access. *IEEE Internet of Things Journal*, 1–1.

Tárnok, A. (2020). Machine learning, COVID-19 (2019-nCoV), and multi-OMICS. *Cytometry, 97*(3), 215.

Tuli, S., Tuli, S., Tuli, R., & Gill, S. S. (2020). Predicting the growth and trend of COVID-19 pandemic using machine learning and cloud computing. *Internet of Things, 11*, 100222.

Vaid, S., Cakan, C. and Bhandari, M. (2020). Using machine learning to estimate unobserved COVID-19 infections in North America. *The Journal of bone and joint surgery, American volume.*

Wang, P., Zheng, X., Li, J., & Zhu, B. (2020). Prediction of epidemic trends in COVID-19 with logistic model and machine learning technics. *Chaos, Solitons & Fractals, 139*, 110058.

Werkman, M., Green, D. M., Murray, A. G., & Turnbull, J. F. (2011). The effectiveness of fallowing strategies in disease control in salmon aquaculture assessed with an SIS model. *Preventive Veterinary Medicine, 98*(1), 64–73.

Who.int. 2021. Weekly epidemiological update-8 December 2020. [online] Available at: <https://www.who.int/publications/m/item/weekly-epidemiological-update-8-december-2020> [Accessed 14 September, 2021].

Wong, H. Y. F., Lam, H. Y. S., Fong, A. H. T., Leung, S. T., Chin, T. W. Y., Lo, C. S. Y., Lui, M. M. S., Lee, J. C. Y., Chiu, K. W. H., Chung, T. W. H., & Lee, E. Y. P. (2020). Frequency and distribution of chest radiographic findings in patients positive for COVID-19. *Radiology, 296*(2), E72–E78.

Xue, J., Chen, J., Hu, R., Chen, C., Zheng, C., Su, Y., & Zhu, T. (2020). Twitter discussions and emotions about the COVID-19 pandemic: Machine learning approach. *Journal of Medical Internet Research, 22*(11), e20550.

Yadav, M., Perumal, M., & Srinivas, M. (2020). Analysis on novel coronavirus (COVID-19) using machine learning methods. *Chaos, Solitons & Fractals, 139*, 110050.

Yan, L., Zhang, H., Goncalves, J., Xiao, Y., Wang, M., Guo, Y., Sun, C., Tang, X., Jin, L., Zhang, M., & Huang, X. (2020). A machine learning-based model for survival prediction in patients with severe COVID-19 infection.

Yang, Z., Zeng, Z., Wang, K., Wong, S. S., Liang, W., Zanin, M., Liu, P., Cao, X., Gao, Z., Mai, Z., & Liang, J. (2020). Modified SEIR and AI prediction of the epidemics trend of COVID-19 in China under public health interventions. *Journal of Thoracic Disease, 12*(3), 165.

Zu, Z. Y., Jiang, M. D., Xu, P. P., Chen, W., Ni, Q. Q., Lu, G. M., & Zhang, L. J. (2020). Coronavirus disease 2019 (COVID-19): A perspective from China. *Radiology, 296*(2), E15–E25.

Vaccination, Lock-Down, Measures and Time-Varying Reproduction Number Based Estimation and Control of COVID-19 Outbreak in Turkey

Meric Cetin and Selami Beyhan

Abstract The COVID-19 outbreak continues to spread all over the world. In many countries, measures are being taken and vaccination studies continue due to the spread of different virus mutations. In order to facilitate the fight against the virus, the data collected from the beginning of the epidemic until today have been used to construct a mathematical model of the epidemic spread. In this study, it was aimed to introduce an extended version of a recent epidemic model that has been proposed, in which parameters such as vaccination rate, quarantine conditions and time-varying virus reproduction number are included. For in case of the reproduction number and some of the states are time-varying and difficult to measure, they are estimated by a recently introduced efficient nonlinear observer. The real-data validation of the estimations shows the accuracy of the mathematical model and indicates that the spread of the virus can be controlled by observer-based control. Therefore, a nonlinear model-predictive control is used to stabilize the dynamics of the outbreak while unmeasured dynamics and parameters are estimated by the designed observer. Numerical results were obtained for the daily vaccinated population of Turkey is 2.1 million individuals ($w = 0.025$) and 420 thousand individuals ($w = 0.005$), respectively. For $w = 0.025$, it is seen that the COVID-19 pandemic will completely terminate in nearly 600 days for the BNT162b2 vaccine and nearly 700 days for the CoronaVac vaccine. For $w = 0.005$, it can be said that herd immunity will be reached on average 1000 days after (March 2020) the first active case is reported in Turkey. In the light of the numerical results, depending on the different quarantine conditions and vaccination rates, the future course of the epidemic in Turkey can be determined. Finally, observer-based stabilization results can be used in other countries where similar measures and vaccination strategies are applied.

M. Cetin (✉)
Pamukkale University, Denizli, Turkey
e-mail: mcetin@pau.edu.tr

S. Beyhan
İzmir Demokrasi University, İzmir, Turkey
e-mail: selami.beyhan@idu.edu.tr

V. Chang et al. (eds.), *Artificial Intelligence and Machine Learning Methods in COVID-19 and Related Health Diseases*, Studies in Computational Intelligence 1023,
https://doi.org/10.1007/978-3-031-04597-4_7

Keywords COVID-19 · Time-varying reproduction number · Vaccination · Predictive control

1 Introduction

Sars-Corona Virus (SARS-CoV), a highly infectious RNA virus, was first seen in China in February 2003 (Wan et al., 2020). SARS-CoV is transmitted between individuals through the respiratory tract, body fluids, or contact and it causes high fever, cough, muscle pain, diarrhea, difficult breathing, colds, pneumonia, and several respiratory infections (Al-Hazmi, 2016). Another Coronavirus infection is the MERS-CoV infection that was first seen in Jordan in 2012 (De Groot et al., 2013). The most important symptoms of MERS-CoV transmitted between individuals through close contact or droplets are fever, cough, shortness of breath, diarrhea, nausea, pneumonia, and vomiting. At the end of December 2019, it was defined that a new Coronavirus infection was encountered with the report of the increase of unusual pneumonia cases to the World Health Organization (WHO) in Wuhan, China's Hubei province (Lai et al., 2020). SARS-CoV2, an enveloped RNA virus identified by zoonotic infection, causes severe respiratory syndrome such as SARS-CoV and MERS-CoV (Rothan and Byrareddy, 2020). After WHO announced the identification of a new virus belonging to the Corona family, the epidemic started to spread rapidly around the world and the pandemic was officially declared on March 11, 2020. Similar to the SARS-CoV and MERS-CoV, the most common symptoms in COVID-19 infection are fever, cough, fatigue, sputum production, headache, hemoptysis, diarrhoea, dyspnoea, extreme tiredness, loss of taste and smell, loss of speech/movement, and lymphopenia (Huang et al., 2020). Comprehensive measures are needed to control such a devastating epidemic and reduce the person-to-person transmission of COVID-19. Many scientists are still investigating COVID-19 infection to develop fast-acting therapeutics and testing potential vaccines.

Dealing with a global pandemic depends on many parameters. Although countries have obligations such as taking economic measures, maintaining education, preventing the spread of infection, testing and providing treatment opportunities in the pandemic, it is also known that some countries do not have the infrastructure to manage such a global epidemic (Abdel-Basset et al., 2021). Therefore, in order to manage the epidemic rationally, each country should take measures that include economic, political and social factors in accordance with its dynamics (Alankar et al., 2021). It is very important to know the mathematical model of an infection in order to interpret various dynamics in short-term or long-term pandemic management. By using disease knowledge, mathematical models that predict the treatment process, mortality rates, and transmission control help identify possible intervention strategies. There are many mathematical models in the literature that examine the dynamics of epidemic diseases. Some of the many papers investigating the COVID-19 pandemic with these models can be summarized as follows.

In (Chen et al., 2020), a time-dependent SIR (susceptible:S, infected:I, and recovered:R) model has been proposed that predicts the COVID-19 trend by dynamically adjusting transmission rate and recovering rate. In the early stages of the outbreak, COVID-19 dynamics were predicted through a SEIR (exposed:E) model based on the epidemiological characteristics of individuals, the clinical progression of the infection and the intervention measures of the authorities (Hou et al., 2020). Viguerie et al. proposed a SEIRD model based on partial differential equations combined with a heterogeneous diffusion model that examines the dynamics of the COVID-19 outbreak (Viguerie et al., 2021). Ivorra et al. have proposed a novel θ-SEIHRD (hospitalized:H, dead:D) model (Ivorra et al., 2020) that addresses the fraction of detected cases according to the actual total infected cases, unlike the SIR, SEIR, or other general-purpose models. SIDARTHE (diagnosed:D, ailing:A, recognized:R, threatened:T, healed:H and extinct:E) models often allow multiple strategies to be implemented to manage the COVID-19 pandemic as in (Higazy, 2020).

There are still limited COVID-19 models that include vaccination dynamics and lockdown conditions. However, an effective vaccination is considered the best way to mitigate a global epidemic such as COVID-19. Although vaccine development is a long time-consuming process, every vaccine with a certain degree of protection becomes important in combating such a global epidemic. Therefore, it would be useful to include vaccine dynamics into the epidemic models to assess the potential impact of COVID-19 vaccines on the population. In some epidemic models, the dynamics of the disease are adapted to new situations with the contribution of time-varying parameters. In Yildiz et al., (2021), the dynamics of the COVID-19 epidemic model in Turkey have been estimated using a time-varying reproduction number. In addition to time-varying parameters, vaccine dynamics were also added to the proposed epidemic model to measure vaccination rate and vaccine efficacy.

To control an epidemic system in the presence of time-varying parameters, a controller based on observer design is needed. There are various linear and nonlinear observers and observer-based control applications to predict the dynamics/parameters of the COVID-19 epidemic models. An EKF-based nonlinear controller (Rajaei et al., 2021) was proposed to manage COVID-19 outbreaks with unmeasurable variables. In Castaños and Mondié (2021), an observer was proposed for a SIR epidemic model that compensates time delays in input and output, and the observer performance is evaluated with the optimal control. An observer-based model predictive control approach for constrained control of the dynamic model of COVID-19 was proposed in Péni et al., (2020). Mathematical models that can determine future dynamics such as treatment process, mortality rates, intensive care bed number monitoring, transmission control provide authorities with short and long-term possible intervention strategies in pandemic management. In this way, simulations such as vaccination strategies, lockdown applications, evaluation of the drug therapy effects can be tested over different epidemic scenarios.

In this paper, following the recent mathematical model and using the data reported by the Republic of Turkey, Ministry of Health (Roser et al., 2020), an extended version of the epidemic model has been proposed (Yildiz et al., 2021) to include the vaccination, quarantine measures and reproduction number. The most important parameter of the outbreak is the reproduction number of the virus, which is time-varying based on the vaccination and lockdown rate, is investigated in detail since its value is directly related to the increase or decrease of the outbreak. An observer-based controller is designed to stabilize the outbreak dynamics where it produces suitable control inputs in terms of vaccine number and measures ratio where some of the states and parameters are estimated using a nonlinear observer. The numerical results will represent the possible outcomes of the governmental precautions with respect to applied vaccination rate and quarantine measures. The estimation results of the nonlinear observer are validated based on the previously collected data; therefore, the model and obtained control results can be used to obtain possible measures in the future. The epidemiological data related to Turkey in this paper cover the time period from March 11, 2020, to April 12, 2021.

2 SARS-CoV-2 Population Dynamics

2.1 Model Description

The most important factor in the transmission of COVID-19 is the number of contacts of infected people and the virus load released by this contact. In addition, inhaling droplets containing the virus or touching the surfaces contaminated with SARS-CoV-2 and then touching the hands to the mouth, nose, or eyes causes the transmission of COVID-19. Along with the social distance rule, wearing face masks correctly is an important measure that prevents individuals from breathing virus-carrying droplets. It is effective to take individual hygiene measures (washing hands frequently with soap, wearing glasses, using disinfectants) to reduce infection through surface contamination. Besides, the lockdown precaution is required to reduce large contamination that would affect the population. While lockdown and social distance practices reduce the rate of contact with the virus, the use of face masks and hygiene measures reduce the possibility of virus transmission (Mondal et al., 2020). In order to control the COVID-19 outbreak in Turkey, as in many other countries, the closure of educational institutions and restaurants, the cancellation of sporting events and cultural and art activities, restrictions on public transport and so on measures have been implemented to achieve a nationwide lockdown. Vaccines being developed alongside existing vaccines hold promise for disease termination. Unlike the individual hygiene measures taken, full or partial lockdown strategies at certain intervals and vaccination, there is no drug that has completed phase 1–3 trials that can cure the SARS-CoV-2 infection. Although the development of a drug requires time, it is thought that drug usage will be a more precise solution in the treatment when the process is completed. For this reason, special drug investigations that can be taken orally or administered intranasally are still ongoing. Until the drug development process is completed, drugs that have completed their safety tests and developed for other purposes can be preferred to cure the infection. In Yildiz et al., (2021), an epidemic model representing the COVID-19 outbreak in Turkey using data reported by the Republic of Turkey, Ministry of Health (Roser et al., 2020) has been proposed to handle the present strategies. In order to analyze to what extent all these restrictions, personal precautions, vaccine strategies, or drug therapies could alter the COVID-19 pandemic, it is necessary to evaluate a computational model. In this paper, we present an extended version of the model (Yildiz et al., 2021). In order that the proposed modified model (1) has been extended to include various control strategies representing lockdown, personal hygiene, use of face masks, social distancing and drug therapies to reduce COVID-19 transmission and, therefore, deaths.

$$
\begin{aligned}
\dot{S}(t) &= \beta_1(t)\tfrac{(1-u_1(t))S(t)I_1(t)}{N} - wS(t),\\
\dot{V}(t) &= wS(t) - q\beta_1(t)\tfrac{(1-u_2(t))V(t)I_1(t)}{N},\\
\dot{E}(t) &= q\beta_1(t)\tfrac{(1-u_2(t))V(t)I_1(t)u_3(t)}{N} + \beta_1(t)\tfrac{(1-u_1(t))S(t)I_1(t)u_3(t)}{N} - p_0E(t),\\
\dot{I}_1(t) &= p_0E(t) - (\mu_1(1-u_3(t)) + \gamma_1 + p_1)I_1(t),\\
\dot{I}_2(t) &= p_1I_1(t)u_3(t) - (\mu_2(1-u_3(t)) + \gamma_2 + p_2)I_2(t),\\
\dot{I}_3(t) &= p_2I_2(t)u_3(t) - (\mu_3 + \gamma_3)I_3(t),\\
\dot{R}(t) &= \gamma_1I_1(t)(1-u_3(t)) + \gamma_2I_2(t)(1-u_3(t)) + \gamma_3I_3(t),\\
\beta_1(t) &= 1.29R_t^2(\mu_1 + \gamma_1 + p_1), 0 \le u_1 \le 1, 0 \le u_2 \le 1, 0 \le u_3 \le 1
\end{aligned}
\tag{1}
$$

The individuals related to the variables (1) showing the population of Turkey at time t is expressed as: S : susceptible, V : vaccinated, E : exposed, I_1 : mildly infectious (mildly infected symptomatic individuals including the ones staying in a health care facility but not in an intensive care unit (ICU)), I_2 : patients staying in the ICU, I_3 : ventilated and R : recovered. In that compartmental model designed to explain the dynamics of the COVID-19 outbreak, $N = S+E+I_1+I_2+I_3+R$ refers to the population. I_1 refers to mildly infected symptomatic individuals who remain in a health facility but are not in an intensive care unit (ICU). $u_1(t), u_2(t), u_3(t)$ are the control signals. $u_1(t)$ represents full or partial lockdown that is considered to reduce the number of contacts. $u_2(t)$ defines personal hygiene, use of face masks, and social distance, which are thought to reduce the rate of contamination. $u_3(t)$ denotes the drug therapy considered to reduce mortality and increase the rate of recovery. In the proposed model (1), it is assumed that all control signals are non-zero values at the same time. In order to control the epidemic, it will be possible to evaluate scenarios in which at least one measure is active. The constant values of some parameters in the SARS-CoV-2 epidemic model are given in Table 1 Symptomatic daily incidence, death, ICU prevalence and ventilator unit prevalence are defined by $y_1(t) = p_0E(t)$, $y_2(t) = \mu_1I_1(t) + \mu_2I_2(t) + \mu_3I_3(t)$, $y_3(t) = I_2(t)$ and $y_4(t) = I_3(t)$, respectively.

Table 1 Parameters used in the SARS-CoV-2 epidemic model

Parameter	Description	Value (day^{-1})
γ_1	Recovery rate for I_1	0.0577075
γ_2	Recovery rate for I_2	0.161953
γ_3	Recovery rate for I_3	0.235026
p_1	Rate of progression from I_1 to I_2	0.0167814
p_2	Rate of progression from I_2 to I_3	0.152005
μ_1	Death rate for I_1	0.00204243
μ_2	Death rate for I_2	0.000221077
μ_3	Death rate for I_3	0.0372249
β_1	Initial value of the contact rate	0.925267
R_0	Initial value of the basic reproduction number	3.477073
p_0	Rate of progression from E to I_1	0.0745585

2.2 *Time-Varying Parameter-Based Model Validation*

The measures taken by the authorities during the spread of an epidemic cause some parameters to change over time. We cannot expect the fixed values of such parameters to be suitable for the pandemic model. In Eq. (1), w, q denote the vaccination rate of susceptible individuals and the efficacy of the vaccination, respectively. $\beta_1(t)$ refers to contact rate, which is based on time-varying reproduction number (R_t). The initial value of the basic reproduction number (R_0) was defined as $R_0 = \sqrt{\beta_1(0)/(\mu_1 + \gamma_1 + p_1)}$ in (Yildiz et al., 2021). R_0 is a mathematical term used to describe the progression of a pathogen in the population during an epidemic disease (Zhao et al., 2020). The magnitude of R_0 refers to the infectiousness of the infection, such that R_t that constantly decreases over time and remains below "1" indicates controlled progression. Many factors such as epidemiological characteristics (transmission mode, incubation period, infectious period), biological characteristics of an infectious agent (infectiousness, virulence, pathogenic), or socio-demographic characteristics affect the value of R_0. According to the literature, the R_0 value of the SARS CoV-2 virus that caused the COVID-19 outbreak is approximately 2.8 (Gul et al., 2020). It is also known that R_0 will increase exponentially as the cycle of transmission increases. If there are susceptible individuals in the population due to vaccination or antiviral drug therapy, more complex estimation methods are needed to estimate R_t. Although it is difficult to estimate the R_t, developing strategies over an effective reproduction number will give more realistic results in alleviating the epidemic burden (Koyama et al., 2021). Based on the data specified in the 2019 Health statistics yearbook, it has stated that the total number of intensive care beds for adults is 39955 and the number of ventilators is 31733. Considering that 75% of these beds and devices are used in healthcare institutions, it will be necessary to realize rational strategies to overcome the pandemic burden through an appropriate pandemic model.

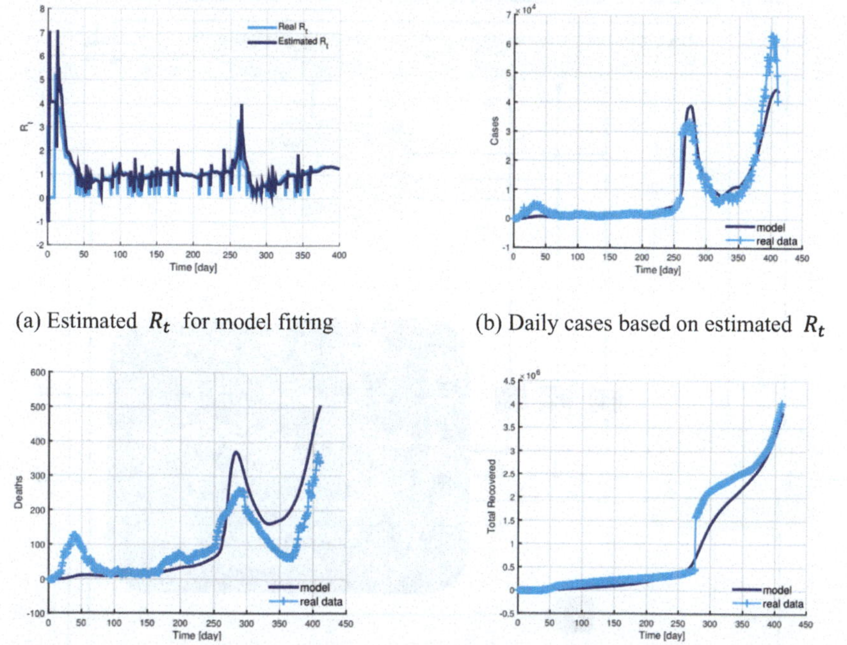

(a) Estimated R_t for model fitting
(b) Daily cases based on estimated R_t
(c) Daily deaths based on estimated R_t
(d) Daily recovered based on estimated R_t

Fig. 1 Case, death and recovered numbers based on estimated R_t for model fitting

In this study, various strategies have been examined through a model based on effective R_t to reduce COVID-19 transmission and, therefore, deaths, manage the usage of intensive care beds and ventilators, and keep the pandemic under control without disrupting healthcare services. For this purpose, a parameter estimation method that predicts $\beta_1(t)$ value has been designed to evaluate model fitting. The daily case, death and recovered numbers for model fitting calculated through the model in (Yildiz et al., 2021) and the estimated R_t from March 11, 2020, to April 12, 2021 are presented in Fig. 1.

3 Observer-Based Control of COVID-19 Outbreak

In this section, an observer-based Nonlinear Model-Predictive Control (NMPC) structure that includes various control strategies such as lockdown, personal hygiene, use of face masks, social distancing and drug therapies to reduce COVID-19 transmission and, therefore, deaths in Turkey is presented. The proposed observer-based NMPC structure is illustrated in Fig. 2. This structure includes the proposed SARS-CoV-2 population model, second-order Gradient-based nonlinear observer block,

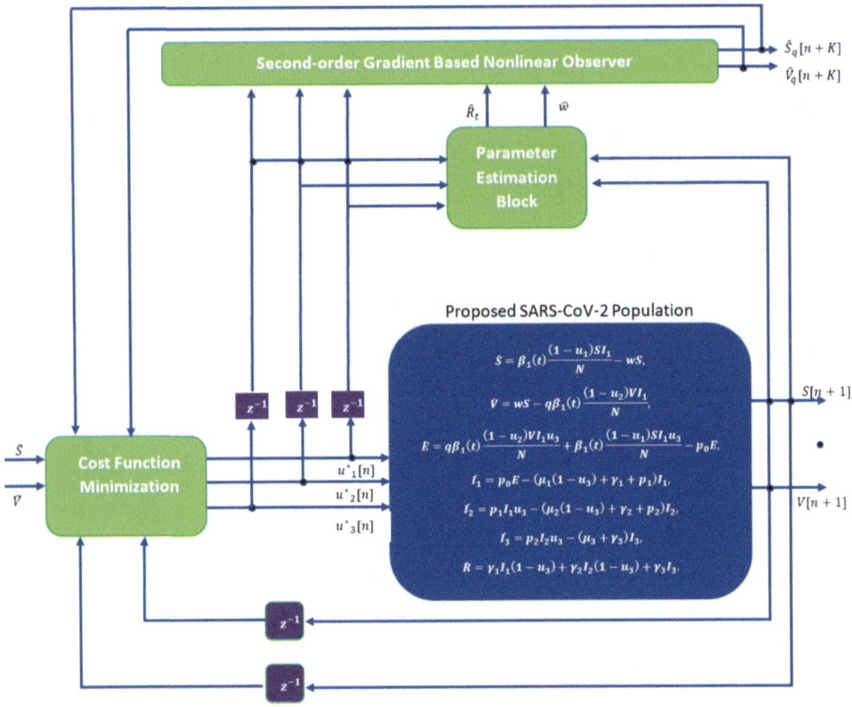

Fig. 2 The block diagram of observer-based NMPC for COVID-19 in Turkey

parameter estimation block and NMPC based Cost Function Minimization (CFM) block. In this control mechanism, the discretized model of the system is responsible for producing $\kappa-$ahead future predictions of the SARS-CoV-2 population dynamics using the candidate control signals and parameter estimation during each sampling period. The main problem for the controller is to find the optimal values of the time-varying reproduction number and the vaccination rate in the parameter vector. Then, the nonlinear system is adaptively controlled with NMPC-based strategies (the combinations of lockdown, personal hygiene, use of face masks, social distancing and drug therapies) produced using these future predictions.

3.1 Nonlinear Model Predictive Control

NMPC, which provides optimization-based predictive generation of control signal under different circumstances, is an elegant controller suitable for constrained, multi-variable problems. The purpose of NMPC is to calculate the optimal control signal along a prediction horizon to minimize an objective function that includes the estimated future outputs of the system in each iteration. For n th-order multi-input multi-output nonlinear discretized system, constrained NMPC is designed as

$$\begin{aligned}
&\min_{u} F(\mathbf{x}[n], \mathbf{u}[n], \boldsymbol{\theta}) \text{subjectto} \\
&\mathrm{X}_i = \{x_i \in \Re | x_{i_{min}} \le x_i \le x_{i_{max}}, i = 1, \dots, n_x\} \\
&\mathrm{U}_r = \{u_r \in \Re | u_{r_{min}} \le u_r \le u_{r_{max}}, r = 1, \dots, n_u\} \\
&\Delta\mathrm{U}_r = \{\Delta u_r \in \Re || \Delta u_r| \le \Delta u_{r_{max}}, r = 1, \dots, n_{ud}\} \\
&\mathrm{Y}_q = \{y_q \in \Re | y_{q_{min}} \le y_q \le y_{q_{max}}, q = 1, \dots, n_y\}
\end{aligned} \tag{2}$$

with that cost function

$$F(\mathbf{u}[n]) = \frac{1}{2} \sum_{p=1}^{\kappa} \sum_{j=1}^{n_y} (\tilde{y}_{j|[n+p]} - \widehat{y}_{j|[n+p]})^2 + \frac{1}{2}\lambda_r \sum_{r=1}^{n_{ud}} (u_{r|[n+1]} - u_{r|[n]})^2, \tag{3}$$

$\mathbf{x} \in \Re^N$, $\mathbf{u} \in \Re^R$, $\mathbf{y} \in \Re^Q$ and $\boldsymbol{\theta}$ denote the SARS-CoV-2 state vector, control strategy vector, output measurements (susceptible and vaccinated individuals) and unknown parameters (R_t, w), respectively. λ_r is the penalty term that prevents undesired changes on the input signals, and κ is the prediction horizon of the output. $\widehat{y}_{q|[n+\kappa]}$ is the estimated output using the discretized model for the qth sample and $\tilde{\mathbf{y}}$ is the reference signal. Furthermore, there are some characteristics in (2), such as input/input slew constraints that occur due to actuator limitations and state/output constraints related to operational limitations. The cost function to be minimized in terms of outbreak management includes suppression of the SARS-CoV-2 dynamics and control strategies used to suppress these dynamics. The constraints to be satisfied include reducing deaths, managing the use of intensive care beds and ventilators, and producing strategies based on effective time-varying reproductive numbers to control the pandemic without disrupting healthcare. The minimization of the constrained cost function is performed by Levenberg–Marquardt. Output predictions are generated when the first component of $\mathbf{u}[n+1] = [u_1[n], \dots, u_r[n]]^{\mathrm{T}}$ is applied to the system as much as κ-times. When $F(\mathbf{u}[n+1]+\delta\mathbf{u}[n+1]) < F(\mathbf{u}[n+1])$ is accomplished, optimal control signals are updated by adding a correction term. To minimize the cost function using second-order Taylor approximation:

$$\begin{aligned}
F(\mathbf{u}[n+1] + \delta\mathbf{u}[n+1]) \cong F(\mathbf{u}[n+1]) &+ \frac{\partial F(\mathbf{u}[n+1])}{\partial(\mathbf{u}[n+1])}\delta\mathbf{u}[n+1] \\
&+ \frac{1}{2}\frac{\partial^2 F(\mathbf{u}[n+1])}{\partial\mathbf{u}^2[n+1]}(\delta\mathbf{u}[n+1])^2
\end{aligned} \tag{4}$$

where $\frac{\partial F(\mathbf{u}[n+1])}{\partial(\mathbf{u}[n+1])}$ is the gradient vector of the F and $\frac{\partial^2 F(\mathbf{u}[n+1])}{\partial\mathbf{u}^2[n+1]}$ is the Hessian matrix of the F with respect to control signal. The cost function must be differentiated with respect to $\delta\mathbf{u}[n+1]$. From first-order necessary condition $\delta\mathbf{u}[n+1] = -\rho\left(\frac{\frac{\partial F}{\partial\mathbf{u}[n+1]}}{\frac{\partial^2 F}{\partial\mathbf{u}^2[n+1]}}\right)$. ρ is an adjustable parameter, which is chosen from the interval [0, 1]. Jacobian matrix can be derived as $\frac{\partial F}{\partial \mathbf{u}[n]} = 2\mathbf{J}_{\mathbf{u}}^T \hat{\mathbf{e}}$ and $\frac{\partial^2 F}{\partial \mathbf{u}^2[n]} \cong 2\mathbf{J}_{\mathbf{u}}^T \mathbf{J}_{\mathbf{u}}$. Substituting $\Delta \mathbf{u}[n+1] = \mathbf{u}[n+1] - \mathbf{u}[n]$, $\hat{\mathbf{e}}$ is the prediction error vector and $\mathbf{J}_{\mathbf{u}}$ matrix is defined as

$$\hat{\mathbf{e}} = \begin{bmatrix} \tilde{y}_1[n+1] - \hat{y}_1[n+1] \\ \vdots \\ \tilde{y}_q[n+\kappa] - \hat{y}_q[n+\kappa] \\ \sqrt{\lambda_r}\Delta \mathbf{u}[n+1] \end{bmatrix}, \ \mathbf{J}_{\mathbf{u}} = - \begin{bmatrix} \frac{\partial \hat{y}_1[n+1]}{\partial u_1[n+1]} & \frac{\partial \hat{y}_1[n+1]}{\partial u_2[n+1]} & \cdots & \frac{\partial \hat{y}_1[n+1]}{\partial u_r[n+1]} \\ \vdots & & & \\ \frac{\partial \hat{y}_q[n+\kappa]}{\partial u_1[n+1]} & \frac{\partial \hat{y}_q[n+\kappa]}{\partial u_2[n+1]} & \cdots & \frac{\partial \hat{y}_q[n+\kappa]}{\partial u_r[n+1]} \\ \sqrt{\lambda_r} & \sqrt{\lambda_r} & \cdots & \sqrt{\lambda_r} \end{bmatrix}. \quad (5)$$

Each term ($\frac{\partial \hat{y}_q[n+\kappa]}{\partial u_r[n+1]}$) in $\mathbf{J}_{\mathbf{u}}$ matrix is derived from the sampled nonlinear system model. The update rule for $\mathbf{u}[n+1]$ is written as

$$\begin{aligned} \mathbf{u}[n+1] &\Leftarrow \mathbf{u}[n] + \delta \mathbf{u}[n+1], \\ \delta \mathbf{u}[n+1] &= -(\mathbf{J}_{\mathbf{u}}^T \mathbf{J}_{\mathbf{u}} + \mu_u \mathbf{I})^{-1} \mathbf{J}_{\mathbf{u}}^T \hat{\mathbf{e}}. \end{aligned} \quad (6)$$

The change of the control signal can be optimally found using (6). However, the mathematical model of the system or approximate model of the system is needed for the Jacobian calculation. In (Iplikci, 2013), the system model is discretized using Runge–Kutta (RK) 4th-order integration. Based on this model, an exact Jacobian is calculated and the control signal is produced. The states and outputs at $(n+1)th$ sampling index can be written as

$$\begin{aligned} \mathbf{x}[n+1] &= \mathbf{x}[n] + \tfrac{1}{6}(\boldsymbol{k}_1 + 2\boldsymbol{k}_2 + 2\boldsymbol{k}_3 + \boldsymbol{k}_4), \\ \mathbf{y}[n+1] &= \mathbf{h}(\mathbf{x}[n+1], \mathbf{u}[n], \boldsymbol{\theta}). \end{aligned} \quad (7)$$

The $\mathbf{k}_i$ variables with $i = 1, \ldots, 4$ are defined as

$$\begin{aligned} \boldsymbol{k}_1 &= T_s \mathbf{g}(\mathbf{x}[n], \mathbf{u}[n], \boldsymbol{\theta}), \\ \boldsymbol{k}_2 &= T_s \mathbf{g}(\mathbf{x}[n] + 0.5\boldsymbol{k}_1, \mathbf{u}[n], \boldsymbol{\theta}), \\ \boldsymbol{k}_3 &= T_s \mathbf{g}(\mathbf{x}[n] + 0.5\boldsymbol{k}_2, \mathbf{u}[n], \boldsymbol{\theta}), \\ \boldsymbol{k}_4 &= T_s \mathbf{g}(\mathbf{x}[n] + \boldsymbol{k}_3, \mathbf{u}[n], \boldsymbol{\theta}). \end{aligned} \quad (8)$$

where it is referred to as the RK discretized model of the system. $\mathbf{g}(.)$, $\mathbf{h}(.)$ functions are known and continuously differentiable with respect to the control inputs, the state variables and the system parameters. For a Lipschitz nonlinear system, stability of discretization can be shown using a sufficiently small step size. At the same time, the unmeasurable states and the unknown model parameters of the system can be updated based on the second-order gradient descent approach. The future behavior of the reference signals is directly embedded in the parameter change, and then fast parameter adaptations are obtained.

3.2 Gradient-Based Nonlinear Observer Design

In order to estimate unmeasurable states and parameters of the COVID-19 outbreak model, a gradient based nonlinear observer is proposed due to its fast estimation capability. The Eq. (8) is available for state and parameter estimation that is a discretized-model of the continuous-time system. The states and parameters of the corresponding system are estimated through the measurement and estimation errors given as

$$\hat{\mathbf{e}} = \begin{bmatrix} \hat{e}_1 \\ \vdots \\ \hat{e}_N \end{bmatrix} = \begin{bmatrix} x_1[n+1] - \hat{x}_1[n+1] \\ \vdots \\ x_N[n+1] - \hat{x}_N[n+1] \end{bmatrix}_{N\times 1} \tag{9}$$

where the objective function is

$$F(\mathbf{x}[n], \mathbf{u}[n]) = \hat{\mathbf{e}}^T[n+1]\hat{\mathbf{e}}[n+1] = \hat{e}_1^2[n+1] + \cdots + \hat{e}_N^2[n+1]. \tag{10}$$

to be minimized by proper update of states and parameters. Here, the minimization of cost function is accomplished by the Levenberg–Marquardt state estimation update:

$$\hat{\mathbf{x}}[n+1] \Leftarrow \hat{\mathbf{x}}[n] - (J_x^T \mathbf{J}_\mathbf{x})^{-1}\mathbf{J}_\mathbf{x}^T \hat{\mathbf{e}}, \tag{11}$$

where $\mathbf{J}_\mathbf{x}$ is the Jacobian matrix given by $[\mathbf{J}_\mathbf{x}] = [\frac{\partial \hat{\mathbf{e}}}{\partial \hat{\mathbf{x}}}] = -\frac{\partial \hat{\mathbf{y}}[n+1]}{\partial \hat{\mathbf{x}}[n]}$. By using the discretized model (8) of the system $\partial \hat{\mathbf{y}}_i / \partial \hat{\mathbf{x}}_j$'s can be obtained as

$$\frac{\partial \hat{\mathbf{y}}_i[n+1]}{\partial \hat{\mathbf{x}}_j[n]} = \left[\sum_{i=1}^{N} \frac{\partial \mathbf{h}[n+1]}{\partial \hat{x}_i[n+1]} \frac{\partial \hat{\mathbf{x}}_i[n+1]}{\partial \hat{\mathbf{x}}_j[n]} \right] \Bigg|_{\substack{\mathbf{x} = \hat{\mathbf{x}}[n] \\ \mathbf{u} = \mathbf{u}[n]}}, \tag{12}$$

$$\frac{\partial \hat{x}_i[n+1]}{\partial \hat{x}_j[n]} = \frac{\partial \hat{x}_i[n]}{\partial \hat{x}_j[n]} + \frac{1}{6}\left(\frac{\partial \mathbf{k}_1[n]}{\partial \hat{x}_j[n]} + 2\frac{\partial \mathbf{k}_2[n]}{\partial \hat{x}_j[n]} + 2\frac{\partial \mathbf{k}_3[n]}{\partial \hat{x}_j[n]} + \frac{\partial \mathbf{k}_4[n]}{\partial \hat{x}_j[n]} \right). \tag{13}$$

The discretized model-based partial derivative components required in (13) are obtained as in (Iplikci, 2013). In order to estimate the unknown parameters of the $\boldsymbol{\theta}$ vector in (8) by an observer, input and output signals are assumed known at n*th* sample index. First, the state estimation error is calculated based on the estimated parameters. Then, its partial derivative is taken with respect to estimated parameters which will construct the Jacobian matrix for parameter estimations. $\mathbf{J}_\theta$ is written as

$$\mathbf{J}_\theta = \begin{bmatrix} \frac{\partial \hat{e}_1}{\partial \theta} \\ \vdots \\ \frac{\partial \hat{e}_N}{\partial \theta} \end{bmatrix} = - \begin{bmatrix} \frac{\partial \hat{x}_1[n+1]}{\partial \theta} \\ \vdots \\ \frac{\partial \hat{x}_N[n+1]}{\partial \theta} \end{bmatrix}_{N\times P} \tag{14}$$

where P represents the number of unknown parameters. The parameter estimation update rule is defined as

$$\begin{aligned}\theta[n+1] &\Leftarrow \theta[n] + \Delta\theta[n+1]\\ \Delta\theta[n+1] &= -(J_\theta^T J_\theta + \mu_\theta \mathbf{I})^{-1} J_\theta^T \hat{\mathbf{e}}.\end{aligned} \tag{15}$$

Note that **I** is a identity matrix with proper dimension. The main issue is here to calculate the partial derivatives $\frac{\partial \widehat{x}_1[n+1]}{\partial \theta}, \ldots, \frac{\partial \widehat{x}_N[n+1]}{\partial \theta}$ in $\mathbf{J_\theta}$

$$\frac{\partial \widehat{x}_i[n+1]}{\partial \theta} = \frac{1}{6}\left(\frac{\partial \mathbf{k}_1}{\partial \theta} + 2\frac{\partial \mathbf{k}_2}{\partial \theta} + 2\frac{\partial \mathbf{k}_3}{\partial \theta} + \frac{\partial \mathbf{k}_4}{\partial \theta}\right). \tag{16}$$

By using the derivative chain rule, the partial derivative vectors in (16) are derived by the discretized model of the nonlinear system as in (Iplikci, 2013). When the equations are accurately defined for state and parameter estimation of a nonlinear system, then the estimation errors converge to zero very fast in the second-order gradient level. The power of the designed observer arises with Jacobian derivation since there is no loss of information about gradient derivations. In the next section, a discretization-based nonlinear observer is used to estimate unmeasurable states and parameters of the COVID-19 model such that state and parameter estimation based NMPC stabilization results will be given as a prominent contribution of the study.

4 Experimental Results

In this section, an observer-based nonlinear model predictive controller has been designed to interpret the COVID-19 infection in Turkey. First, the designed observer was used to estimate the vaccination rate (w) and the time-varying reproduction number (R_t). Then, the parameter estimation process was carried out in the presence of various restrictions, individual precautions, vaccination contributions, or drug treatments. In the proposed model (1), it is assumed that all control signals are non-zero values simultaneously. It will be possible to evaluate scenarios in which at least one measure is active to control the epidemic.

4.1 *Time-Varying Reproduction Number Scenarios and Estimation Results*

In order to evaluate the situation of the pandemic after April 12, 2021, the effect of the time-varying R_t in (1) was examined over four different scenarios. All states of SARS CoV-2 and the time-varying parameter (R_t) are simultaneously estimated when the reproduction number changes as sinusoidal, exponential, constant, and sawtooth time series. The estimated dynamics of the SARS CoV-2 in (1) without control strategies for time-varying sinusoidal R_t scenarios are shown in Fig. 3. In simulations, the design parameters of the observer are chosen as $\mu_\theta = 10^3$ and $\mu_x = 10^{-7}$, respectively. The initial values of the SARS CoV-2 dynamics and R_t are selected as zero. The duration of the simulation is 1200 days and the sampling period is $T_s = 1$ day. The numerical results of the state and parameter estimation for time-varying R_t scenarios are listed in Table 2.

In the process of combating COVID-19, if quarantine measures, drug treatment procedures and individual hygiene measures are not included in the epidemic population dynamics, the spread of the outbreak will increase. According to the obtained

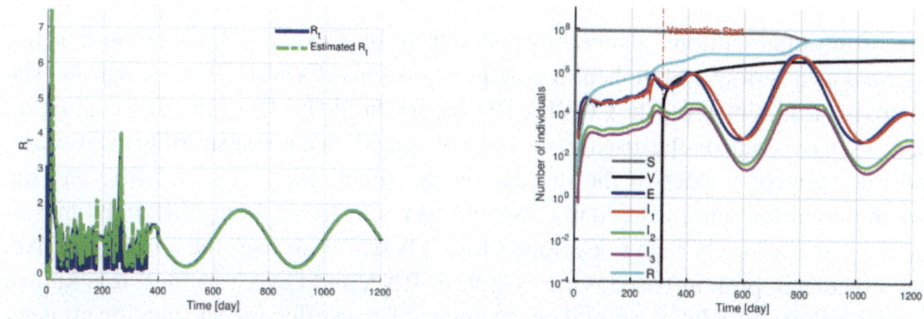

(a) $R_t(t) = 1 - 0.8\sin(0.005\pi t)$ (b) $R_t(t) = 1 - 0.8\sin(0.005\pi t)$

Fig. 3 R_t scenario after April 12, 2021, and SARS CoV-2 dynamics

Table 2 Numerical performances without control strategies

Performance	Vaccination rate	Vaccination rate
$(RMSE(e) = \sqrt{\frac{1}{T}\int_0^T e(t)^2 dt})$	$w = 0.025$	$w = 0.005$
Reproduction number	$R_t(t) = 1 - 0.8\sin(0.005\pi t)$	
Parameter estimation error	0.035891	0.041119
Total state estimation error	5.504951e12	5.551947e12
Reproduction number	$R_t(t) = 0.8e^{(0.0015t)} - 0.5$	
Parameter estimation error	0.178013	0.173295
Total state estimation error	5.362730e12	5.446087e12
Reproduction number	$R_t(t) = 1$	
Parameter estimation error	0.032419	0.038285
Total state estimation error	5.378976e12	5.449229e12
Reproduction number	$R_t(t) = 1.3 - 0.8\text{sawtooth}(8\pi t)$	
Parameter estimation error	0.051717	0.052864
Total state estimation error	5.521170e12	5.501626e12

results, it is shown that ICU (I_2) and ventilator (I_3) capacities are exceeded in all time-varying R_t scenarios. In all scenarios where R_t value is above 1, effective vaccination and control strategies are required in order not to collapse the health system.

4.2 Time-Varying Reproduction Number Based Control Results

4.2.1 The Efficiencies of CoronaVac and BNT162b2 Vaccines

One of the interventions planned to respond to the COVID-19 pandemic that has emerged as a serious public health emergency is mass COVID-19 vaccination. The groups to be administered the COVID-19 vaccine should be determined by evaluating the risks of exposure to the disease, the risks of spreading and transmitting the disease, and the negative impact of the disease on the functioning of social life, and the vaccine should be administered to these groups respectively. Many different types of vaccines are currently being developed for COVID-19 all over the world: inactive, live attenuated, protein-based, viral vector, m-RNA and DNA vaccines. It is known that 75% of the population should be vaccinated for an effective vaccination strategy in the presence of a global epidemic. Assuming that the population of Turkey is approximately 84 million, 63 million people should be vaccinated in order to achieve herd immunity. National vaccination for the COVID-19 pandemic in Turkey started on January 13, 2021. Two different types of vaccination are already applied as of April 2021. The first of these is the CoronaVac (inactive vaccine) developed by Sinovac, and the other is the BNT162b2 (m-RNA and DNA vaccine) developed by Pfizer and BioNTech. The efficiencies of CoronaVac (Palacios et al., 2020; Wu et al., 2021) and Pfizer/BioNTech (Amit et al., 2021; Polack et al., 2020) are approximately 70%, 90%, which correspond to $q = 0.3$ and $q = 0.1$, respectively.

4.2.2 Model Constraints and Control Objectives

In this paper, it was aimed to control the vaccine and susceptible dynamics under multi-control strategy by considering both types of vaccines while designing an observer-based NMPC. In this case, the convergence of susceptible individuals to zero simultaneously in a period that included the two doses vaccination of approximately 63 million people was examined. The reference signal is defined for the number of individuals to be vaccinated as 63 million, and the reference signal is set to zero for the complete termination of susceptible individuals. According to the population of Turkey, the daily amount of vaccination is thought to be 2.1 million individuals and 420 thousand individuals while $w = 0.025$ and $w = 0.005$, respectively. In addition, it was thought that the current intensive care bed capacity and the usage of ventilators in the health infrastructure should be limited to 75%. During the NMPC design process, the maximum and minimum values of these dynamics in the population equations were added as constraints.

In Eq. (1), $u_1(t)$ represents a full or partial lockdown that is considered to reduce the number of contacts. $u_2(t)$ defines personal hygiene, use of face masks, and social distance, which are thought to reduce the rate of contamination. $u_3(t)$ denotes the drug therapy considered to reduce mortality and increase the rate of recovery. The limits of the control signals are $u_{i_{min}} = 0$, $u_{i_{max}} = 1$, $(i = 1, \ldots, 3)$. The input/input slew constraints caused by actuator limitations and status/output constraints related to operational limitations are as: $\Delta u_1 = 0.01$, $\Delta u_2 = 0.02$, $\Delta u_3 = 0.001$. Infection risk may vary for different scenarios of time-varying reproduction numbers. In this case, the control strategies produced by the designed controller differ according to the outbreak spread in the scenarios considered. The change of control signals between 0 and 1 indicates the degree of action to be taken. Small values of the control signals indicate that the measures can be loosened, while larger values indicate that the measures should be applied more stringently. The limits of the patients staying in the ICU are $I_{2_{min}} = 0$, $I_{2_{max}} = 30000$ and the limits of the ventilated individuals are $I_{3_{min}} = 0$, $I_{3_{max}} = 24000$, respectively. The initial conditions of SARS CoV-2 population dynamics are selected as: $S_0 = 8e7$, $V_0 = 0$, $E_0 = 690$, $I_{1,0} = 500$, $I_{1,0} = 10$, $I_{1,0} =$, $R_0 = 0$. The design parameters of the observer-based NMPC are as: $\kappa = 10$, $\mu = 2 \times 10^8$, $\lambda_1 = \lambda_2 = \lambda_3 = 0.01$. Assuming that R_t changes sinusoidally for $w = 0.025$ and $w = 0.005$, logarithmically, all states of SARS CoV-2 and control signals of observer-based NMPC are shown in Fig. 4 and Fig. 5. The numerical performances of the observer-based nonlinear model predictive controller for time-varying R_t scenarios when $q = 0.3$ and $q = 0.1$ are listed in Tables 3 and 4, respectively. According to the numerical results, it is seen that the difference between the efficacy of the vaccine is negligible when the vaccination rate and the transmission rate are low.

5 Discussions and Conclusion

The numerical results indicate that the designed observer-based NMPC provides satisfactory stabilization results of outbreak dynamics even in the presence of time-varying parameters. According to the Republic of Turkey Ministry of Health, Turkey has an established health infrastructure that can vaccinate 2 million people per day in case of available vaccine supplies. Therefore, numerical results are obtained here for the daily vaccinated population of Turkey is 2.1 million individuals ($w = 0.025$) and 420 thousand individuals ($w = 0.005$), respectively. In the first scenario, if the daily vaccination amount can be kept at the level of 2.1 million people, it is seen that the COVID-19 pandemic will completely terminate in nearly 600 days for the BNT162b2 vaccine and nearly 700 days for the CoronaVac vaccine after the first active case was reported in Turkey (March 2020). With this vaccination rate, it is not possible to vaccinate 75% of Turkey's population in a short time. Therefore, when the estimation results of vaccination rate are analyzed based on the date of January 13,

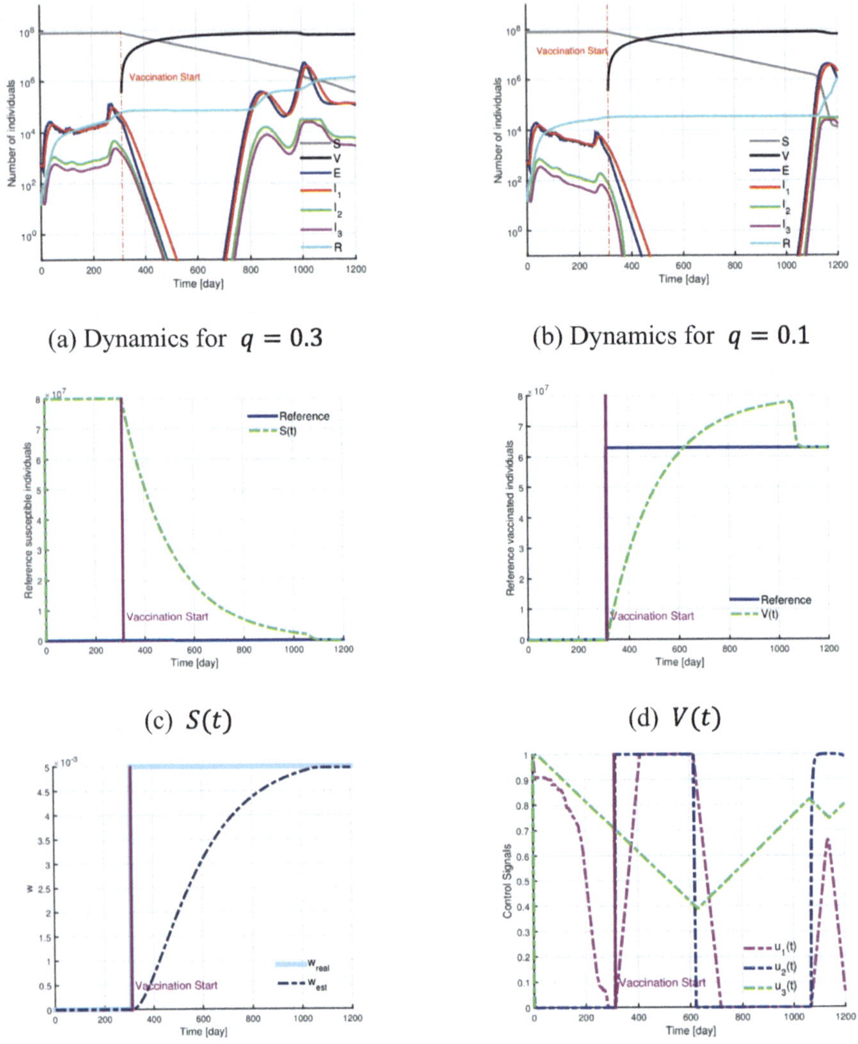

(a) Dynamics for $q = 0.3$ (b) Dynamics for $q = 0.1$

(c) $S(t)$ (d) $V(t)$

(e) w (f) Control signals

Fig. 4 Vaccination strategy with $w = 0.005$ for Turkey: state and parameter estimation results of the SARS CoV-2 under control strategies when $R_t(t) = 1 - 0.8\sin(0.005\pi t)$

2021, it is seen that the desired herd immunity will be reached in June 2021. While the daily vaccination amount is at the level of 2.1 million individuals, social distance, individual hygiene measures (washing hands frequently with soap, wearing glasses, using disinfectants), wearing face masks and drug treatments must continue until June 2023, even if quarantine measures are weakened by the government. According to the second scenario, if the daily vaccination amount can be kept at the level of 420

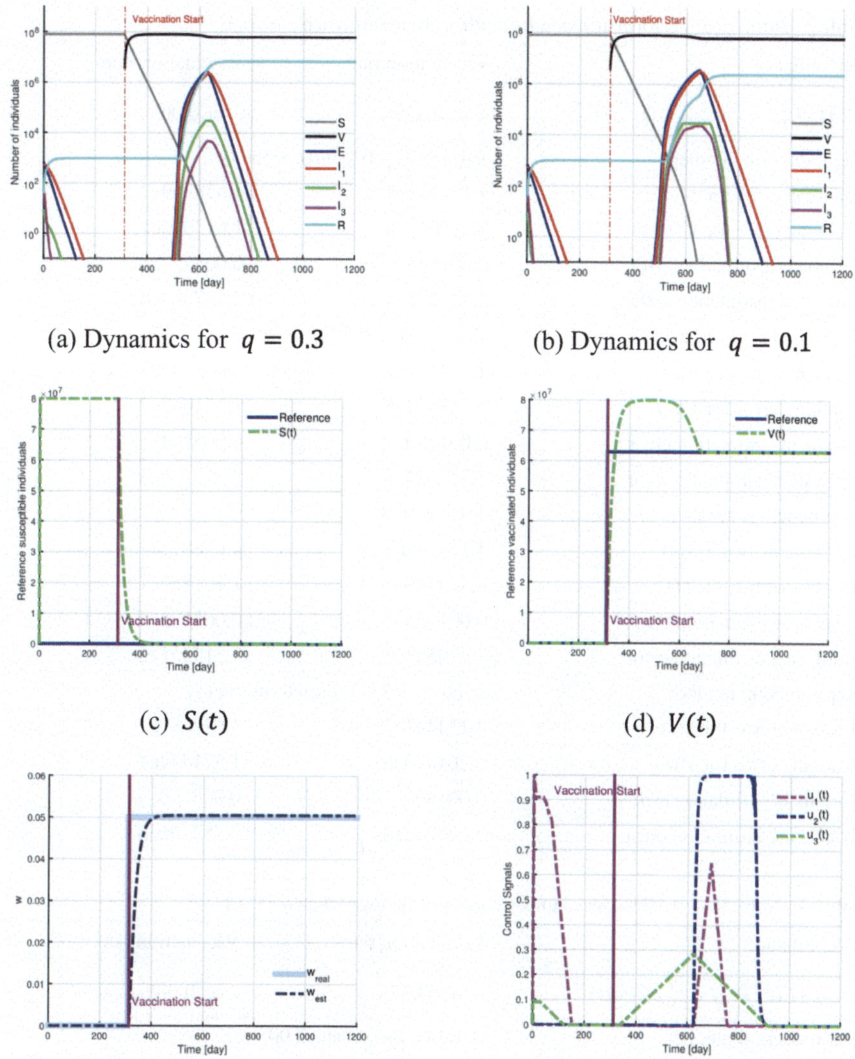

(a) Dynamics for $q = 0.3$

(b) Dynamics for $q = 0.1$

(c) $S(t)$

(d) $V(t)$

(e) w

(f) Control signals

Fig. 5 Vaccination strategy with $w = 0.025$ for Turkey: state and parameter estimation results of the SARS CoV-2 under control strategies when $R_t(t) = 1 - 0.8\sin(0.005\pi t)$

thousand individuals, it can be said that herd immunity will be reached on average 1000 days after (March 2020) the first active case is reported in Turkey. However, even if herd immunity is obtained, an increase has been observed in the number of exposed individuals, mildly infectious individuals, patients staying in the ICU and ventilated individuals. Among the measures included in the fight against COVID-19 are the

Table 3 Numerical Performances under control strategies when $q = 0.3$

Performance	Vaccination rate	Vaccination rate
$(RMSE(e) = \sqrt{\frac{1}{T}\int_0^T e(t)^2 dt})$	$w = 0.025$	$w = 0.005$
Reproduction number	$R_t(t) = 1 - 0.8\sin(0.005\pi t)$	
Tracking error for $S(t)$	4.134444e7	4.678399e7
Tracking error for $V(t)$	7.912612e6	1.613080e7
Parameter estimation error	0.001895	0.005757
Total state estimation error	2.320474e6	2.337783e6
Reproduction number	$R_t(t) = 0.8e^{(0.0015t)} - 0.5$	
Tracking error for $S(t)$	4.134433e7	4.672496e7
Tracking error for $V(t)$	6.467433e6	1.521007e7
Parameter estimation error	0.001780	0.005754
Total state estimation error	2.326022e6	2.357736e6
Reproduction number	$R_t(t) = 1$	
Tracking error for $S(t)$	4.134149e7	4.678878e7
Tracking error for $V(t)$	5.462466e6	1.701559e7
Parameter estimation error	0.001895	0.005672
Total state estimation error	2.495593e6	2.310551e6
Reproduction number	$R_t(t) = 1.3 - 0.8\text{sawtooth}(8\pi t)$	
Tracking error for $S(t)$	4.134444e7	4.677689e7
Tracking error for $V(t)$	6.204813e6	1.571444e7
Parameter estimation error	0.001898	0.005750
Total state estimation error	2.327212e6	2.352706e6

Table 4 Numerical Performances under control strategies when $q = 0.1$

Performance	Vaccination rate	Vaccination rate
$(RMSE(e) = \sqrt{\frac{1}{T}\int_0^T e(t)^2 dt})$	$w = 0.025$	$w = 0.005$
Reproduction number	$R_t(t) = 1 - 0.8\sin(0.005\pi t)$	
Tracking error for $S(t)$	4.134444e7	4.678849e7
Tracking error for $V(t)$	9.120159e6	1.697119e7
Parameter estimation error	0.001895	0.005758
Total state estimation error	2.320605e6	2.312428e6
Reproduction number	$R_t(t) = 0.8e^{(0.0015t)} - 0.5$	

(continued)

Table 4 (continued)

Performance	Vaccination rate	Vaccination rate
Tracking error for $S(t)$	4.134443e7	4.6754884e7
Tracking error for $V(t)$	7.677040e6	1.540282e7
Parameter estimation error	0.001773	0.005757
Total state estimation error	2.312586e6	2.361120e6
Reproduction number	$R_t(t) = 1$	
Tracking error for $S(t)$	4.134168e7	4.678869e7
Tracking error for $V(t)$	6.489067e6	1.714975e7
Parameter estimation error	0.001895	0.005715
Total state estimation error	2.371552e6	2.309524e6
Reproduction number	$R_t(t) = 1.3 - 0.8\text{sawtooth}(8\pi t)$	
Tracking error for $S(t)$	4.134444e7	4.678079e7
Tracking error for $V(t)$	7.834823e6	1.598395e7
Parameter estimation error	0.001852	0.005756
Total state estimation error	2.311211e6	2.345726e6

closure of schools across the country, banning mass gathering events, restrictive measures for public officials, additional travel restrictions, curfews, and lockdown. Since lockdown measures cannot be taken for a long-time due to socio-economic reasons, it can be stated that measures such as social distance, individual hygiene, use of face masks and drug treatments will be tighter and longer for that vaccination rate. In addition, COVID-19 treatment procedures such as starting treatment early in cases with symptoms, taking measures to prevent the development of pneumonia, delaying intensive care and intubation using high flow oxygen therapy should also be continued. The numerical results are producing that how long time is required for the control of the outbreak. We must know that these stabilization results depend on the governmental precautions and vaccine conditions up to today. In case of better conditions exist, the stabilization results will become much faster than the presented here. In addition, proposed observer-based control results can be used to see future dynamics of the outbreak in other countries. Unfortunately, due to the new variants of the virus emerging, vaccination and measures must be continued continuously and systematically to stabilize the virus in the near future.

References

Abdel-Basset, M., Chang, V., & Nabeeh, N. A. (2021). An intelligent framework using disruptive technologies for COVID-19 analysis. *Technological Forecasting and Social Change, 163*, 120431.

Alankar, B., Kaur, H., Ahsaan, S., Sharma, G., & Chang, V. (2021). Towards Reviewing an Immediate Impact of COVID-19 on the Integrative World Economy: An Evolving Perspective. *Journal of Global Information Management (JGIM), 30*(4), 1–19.

Al-Hazmi, A. (2016). Challenges presented by MERS corona virus, and SARS corona virus to global health. *Saudi Journal of Biological Sciences, 23*(4), 507–511.

Amit, S., Regev-Yochay, G., Afek, A., Kreiss, Y., & Leshem, E. (2021). Early rate reductions of SARS-CoV-2 infection and COVID-19 in BNT162b2 vaccine recipients. *The Lancet, 397*(10277), 875–877.

Castaños, F., & Mondié, S. (2021). Observer-based predictor for a susceptible-infectious-recovered model with delays: An optimal-control case study. *International Journal of Robust and Nonlinear Control.*

Chen, Y. C., Lu, P. E., Chang, C. S., & Liu, T. H. (2020). A time-dependent SIR model for COVID-19 with undetectable infected persons. *IEEE Transactions on Network Science and Engineering, 7*(4), 3279–3294.

De Groot, R. J., Baker, S. C., Baric, R. S., Brown, C. S., & Drosten, C., et al. (2013). Commentary: Middle east respiratory syndrome coronavirus (mers-cov): announcement of the coronavirus study group. *Journal of virology, 87*(14), 7790–7792.

Gul, S., Tuncay, K., Binici, B., & Aydin, B. B. (2020). Transmission dynamics of Covid-19 in Italy, Germany and Turkey considering social distancing, testing and quarantine. *The Journal of Infection in Developing Countries, 14*(07), 713–720.

Hitchings, M. D., Ranzani, O. T., Torres, M. S. S., Oliveira, S. B., & Almiron, M., et al. (2021). Effectiveness of CoronaVac in the setting of high SARS-CoV-2 P. 1 variant transmission in Brazil: A test-negative case-control study. *medRxiv*.

Hou, C., Chen, J., Zhou, Y., Hua, L., Yuan, J., et al. (2020). The effectiveness of quarantine of Wuhan city against the Corona Virus Disease 2019 (COVID-19): A well-mixed SEIR model analysis. *Journal of Medical Virology, 92*(7), 841–848.

Huang, C., Wang, Y., Li, X., Ren, L., Zhao, J., et al. (2020). Clinical features of patients infected with 2019 novel coronavirus in Wuhan. *China. the Lancet, 395*(10223), 497–506.

Iplikci, S. (2013). Runge-Kutta model-based adaptive predictive control mechanism for nonlinear processes. *Transactions of the Institute of Measurement and Control, 35*(2), 166–180.

Ivorra, B., Ferrández, M. R., Vela-Pérez, M., & Ramos, A. M. (2020). Mathematical modeling of the spread of the coronavirus disease 2019 (COVID-19) taking into account the undetected infections. The case of China. *Communications in Nonlinear Science and Numerical Simulation, 88*, 105303.

Koyama, S., Horie, T., & Shinomoto, S. (2021). Estimating the time-varying reproduction number of COVID-19 with a state-space method. *PLoS Computational Biology, 17*(1), e1008679.

Lai, C. C., Shih, T. P., Ko, W. C., Tang, H. J., & Hsueh, P. R. (2020). Severe acute respiratory syndrome coronavirus 2 (SARS-CoV-2) and coronavirus disease-2019 (COVID-19): The epidemic and the challenges. *International Journal of Antimicrobial Agents, 55*(3), 105924.

Mondal, C., Adak, D., Majumder, A., & Bairagi, N. (2020). Mitigating the transmission of infection and death due to SARS-CoV-2 through non-pharmaceutical interventions and repurposing drugs. *ISA Transactions.*

Palacios, R., Patiño, E. G., Oliveira Piorelli, R., Conde, M. T. R. P., & Batista, A. P., et al. (2020). Double-blind, randomized, placebo-controlled phase III clinical trial to evaluate the efficacy and safety of treating healthcare professionals with the adsorbed COVID-19 (Inactivated) vaccine manufactured by sinovac–PROFISCOV: A structured summary of a study protocol for a randomised controlled trial. *Trials, 21*(1), 1–3.

Péni, T., Csutak, B., Szederkényi, G., & Röst, G. (2020). Nonlinear model predictive control with logic constraints for COVID-19 management. *Nonlinear Dynamics, 102*(4), 1965–1986.

Polack, F. P., Thomas, S. J., Kitchin, N., Absalon, J., & Gurtman, A., et al. (2020). Safety and efficacy of the BNT162b2 mRNA Covid-19 vaccine. *New England Journal of Medicine.*

Rajaei, A., Raeiszadeh, M., Azimi, V., & Sharifi, M. (2021). State estimation-based control of COVID-19 epidemic before and after vaccine development. *Journal of Process Control, 102*, 1–14.

Roser, M., Ritchie, H., Ortiz-Ospina, E., & Hasell, J. (2020). Coronavirus pandemic (COVID-19). Our world in data.

Rothan, H. A., & Byrareddy, S. N. (2020). The epidemiology and pathogenesis of coronavirus disease (COVID-19) outbreak. *Journal of Autoimmunity, 109*, 102433.
Viguerie, A., Lorenzo, G., Auricchio, F., Baroli, D., Hughes, T. J., et al. (2021). Simulating the spread of COVID-19 via a spatially-resolved susceptible–exposed–infected–recovered–deceased model with heterogeneous diffusion. *Applied Mathematics Letters, 111*, 106617.
Wan, Y., Shang, J., Graham, R., Baric, R. S., & Li, F. (2020). Receptor recognition by the novel coronavirus from Wuhan: An analysis based on decade-long structural studies of SARS coronavirus. *Journal of Virology, 94*(7), e00127-e220.
Wu, Z., Hu, Y., Xu, M., Chen, Z., Yang, W., et al. (2021). Safety, tolerability, and immunogenicity of an inactivated SARS-CoV-2 vaccine (CoronaVac) in healthy adults aged 60 years and older: A randomised, double-blind, placebo-controlled, phase 1/2 clinical trial. *The Lancet Infectious Diseases, 21*(6), 803–812.
Yildiz, T. A., Kose, E., & Tuncer, N. (2021). Assessment of Vaccination and Underreporting on COVID-19 Infections in Turkey Based On Effective Reproduction Number. *medRxiv*.
Zhao, S., Lin, Q., Ran, J., Musa, S. S., Yang, G., et al. (2020). Preliminary estimation of the basic reproduction number of novel coronavirus (2019-nCoV) in China, from 2019 to 2020: A data-driven analysis in the early phase of the outbreak. *International Journal of Infectious Diseases, 92*, 214–217.

Diabetes Analysis with a Dataset Using Machine Learning

Victor Chang, Saiteja Javvaji, Qianwen Ariel Xu, Karl Hall, and Steven Guan

Abstract Diabetes is a disease that actually impacts the capacity of the body to obtain blood glucose, which is usually referred to as blood sugar. At the end of 2019, a new public health problem (COVID-19) emerged. This disease has greatly harmed people with diabetes. Therefore, we intend to make use of data mining algorithms to prevent death and improve the quality of life through the prediction of diabetes. In this paper, four different algorithms have been used to analyze Diabetes from DAT260x Lab01: Logistic, Decision Tree Classifier, Xgboost and SVC. The models are evaluated for which algorithm is much effective. The paper then provides a quick overview of both the set of data and the fieldwork carried out on the subject. In the adjoining step, the dataset and its features are discussed. In addition, the paper explains the four algorithms and virtual environments that have been used to clarify the variables, which have the largest impact on raw data. The findings are obtained by evaluating the confusion matrix applied to the whole selected algorithm. The paper outlines the full observations and conclusions taken based on the results.

Keywords Machine learning · Logistic algorithm · Decision tree classifier · Xgboost algorithm · SVC

V. Chang (✉)
Department of Operations and Information Management, Aston Business School, Aston University, Birmingham, UK
e-mail: v.chang1@aston.ac.uk; victorchang.research@gmail.com

S. Javvaji · Q. A. Xu · K. Hall
Cybersecurity Information Systems and AI Research Group, School of Computing, Engineering and Digital Technologies, Teesside University, Middlesbrough, UK
e-mail: qianwen.ariel.xu@gmail.com; Q.Xu@tees.ac.uk

K. Hall
e-mail: drazarx3@gmail.com; K.Hall@tees.ac.uk

S. Guan
School of Computing, Xian Jiaotong-Liverpool University, Suzhou, China
e-mail: Steven.Guan@xjtlu.edu.cn

V. Chang et al. (eds.), *Artificial Intelligence and Machine Learning Methods in COVID-19 and Related Health Diseases*, Studies in Computational Intelligence 1023,
https://doi.org/10.1007/978-3-031-04597-4_8

1 Introduction

Diabetes is a metabolic disease caused by high sugar levels and is also called blood sugar. Blood glucose is the primary cause of stamina, and it is the result of converting the food consumed. Insulin, a protein secreted by the pancreas, serves to bring glucose from your nutrition towards its muscles such that you can use it for energy. Sometimes, the pancreas may not be able to make sufficient hormones or use hormones well.

Glucose will reside in one's plasma and will not enter its tissue. There are three forms of diabetes. Category 1 diabetes can occur at any time and most commonly occurs in children and adults. (Medicalnewstoday, 2020). When a person has category one diabetes, his or her body produces little or no glucose, which means that the person needs insulin to keep the blood sugar steady every day (Medicalnewstoday, 2020).

Type 2 is more common in teenagers and records about 90% of cases of insulin. When a person has category two diabetes, the glucose generated cannot allow sufficient use of the body. Nonetheless, most people with type 2 diabetes may need an iron supplement or glucose medication to maintain a normal glucose level. It is indeed a type of diabetes that results in higher sugar levels during delivery, but that is consistent with both mother's and baby's complications. GDM completely passes, but women are far more prone to diabetes early in adulthood due to pregnancy (NIDDK, 2020).

The person affected with diabetes should take proper medication and a healthy diet to maintain the diabetes level constant. Moreover, patients should have a blood test for a month to know the glucose level in the blood. Given a diagnosis of type 1 diabetes, the medication drug product needs to last a lifespan. Although type 2 diabetes is a medical illness, antibiotics, especially in the form of tablets, may ultimately be necessary.

Without treatment, diabetes will lead to life-threatening consequences for the nervous, renal, and cardiovascular systems. Moreover, patients who have diabetes are vulnerable to any virus, and they are a high-risk group for the COVID-19. With the support of Artificial Intelligence, including data mining and machine learning, doctors are able to provide an accurate and efficient diagnosis for potential patients, especially the ones with diabetes but not being aware of it.

This chapter aims to employ four machine learning algorithms and evaluate their performance on a public diabetes dataset. The dataset is taken from the Kaggle dataset, which contains 15,001 observations, ten variables. This dataset is used to predict the presence of diabetes in people by using machine learning algorithms, including Logistic, Decision Tree Classifier, Xgboost and SVC and their accuracy values are evaluated to identify the most effective algorithm.

2 Literature Review

2.1 Diabetes and Covid-19

Sars-cov-2 is the cause of coronavirus disease 2019 (COVID-19). On 30 January 2020, the Who declared the outbreak a Public Health Emergency of International Concern, followed by a pandemic crisis on 11 March 2020. According to WHO (2021), by 26 October 2021, the COVID-19 infected 243,572,402 people globally, including 4,948,434 deaths. Generally speaking, people with diabetes may experience severe symptoms and complications when they are sick. If they also suffer from heart disease or other chronic diseases, they are greatly likely to be at higher risk of serious complications when they are infected with COVID-19 (American Diabetes Association, 2021).

Several investigations have shown that "increasing glucose levels increases the speed of SARS-CoV-2 replication." The severity of SARS-CoV-2 infection in diabetic patients is higher than that in non-diabetic patients, and poor blood glucose control indicates an increase in the demand for drugs, hospitalization, and mortality (Lim et al., 2021). The immediate cause of death in some patients who died from the virus was not the virus but diabetes. Medications of COVID-19 can impair the blood sugar level control of the COVID-19 patient with diabetes. The use of glucocorticoids will raise blood glucose, hypertonic dehydration, ketoacidosis, and even lead to life-threatening conditions (Deng et al., 2020). Diabetes and symptoms of viral infection influence and aggravate each other, resulting in further deterioration of the condition (Wu et al., 2020; Yang et al., 2010).

In addition, Covid19 has also had a significant impact on the self-management of diabetic patients. According to Hartmann-Boyce et al. (2020), Covid-19 destroys health care services and interrupts access to drugs and supplies, which may lead to worse outcomes for diabetic patients. Physical activities and diet are the main means of self-management of diabetes, which can reduce the risk of worse outcomes in patients with diabetes and those with cardiac metabolic diseases. However, the current epidemic and social isolation have increased rates of anxiety and depression, making the patient's diet less healthy and exercise less. To make matters worse, half of the people with diabetes do not realize their conditions (Chaves & Marques, 2021), and their situation is not monitored and controlled. If they are infected with coronavirus, there may encounter more severe symptoms and complications. Therefore, it is important to know the patient's diabetes history when treating a patient with COVID-19, as this will help the doctor or specialist make the most appropriate diagnosis.

Artificial Intelligence methods use emerging mathematical algorithms to collect, analyze massive data, and then use the results to predict future events and search for patterns and trends in them. With medical data and advanced techniques, professionals are able to diagnose disease much more quickly and accurately (Chitra & Seenivasagam, 2013).

2.2 Diabetes Prediction and AI Techniques

Numerous research has attempted to deal with the issue of diabetes prediction with the support of machine learning or artificial intelligence techniques, including support vector machine (SVM), Decision Tree, Gradient Boosting Decision Tree, Naive Bayes, ANN (Artificial Neural Network), etc.

The early diagnosis of diabetes is the focus of the studies on diabetes prediction as early diabetes. Suitable treatment can largely minimize the expenditure and mortality in the subsequent phases. Fitriyani et al. (2019) developed a disease prediction model for the early diagnosis of hypertension and diabetes. Their model employed iForest to remove outliers, SMOTETomek to balance data distribution and an ensemble learning approach to make the prediction, and it was proved to have high accuracy. Samant and Agarwal (2018a, 2018b) and Rashid et al. (2021) carried out a comparative analysis on the performance of Random Forest with other ML algorithms. In their research, Random Forest showed better ability in making care decisions. A number of other studies have also contributed to this area, but with different ML algorithms, for example, Convolutional Neural Network (CNN) combined with Long Short-Term Memory (LSTM) (Swapna et al., 2018), Support Vector Machine (SVM) (Komi et al., 2017; Sisodia & Sisodia, 2018), CatBoost (Kumar et al., 2021), etc.

Based on the prior literature, our research selects four different algorithms, Logistic, Decision Tree Classifier, Xgboost and SVC, to analyze the Diabetes dataset from DAT260x Lab01 from Kaggle.

2.3 Logistic Regression

Logistic Regression (LR) is one of the generalized linear models in which the dependent variable is binary, which means that the output of dependent variables can only be two values, "0" and "1" (Christodoulou et al., 2019). In different cases, the meanings of the two values can be healthy/ill, true/false, succeed/fail, etc. LR has many similarities with multiple linear regression analyses. The biggest difference is that LR analyses the relationship between the probability of the dependent variable taking a certain value and the independent variable. In contrast, linear regression directly analyses the relationship between the dependent variable and the independent variable. LR is used in a number of areas, including medical science and social science. In order to assess the severity of patients, LR has been used to develop several medical scales, such as the Trauma and Injury Severity Score (TRISS) (Boyd et al., 1987; Kologlu et al., 2001). Moreover, LR can be employed in engineering, particularly to predict the probabilistic of a certain product, system, or process. Companies employ the technique to predict the tendency of a consumer to buy a product or stop a subscription. In economics, it can be used to predict the probability that a person chooses to enter the labor market.

2.4 *Decision Tree*

The Decision Tree (DT) is a tree that sorts instances based on feature values to classify them. Each node of the DT stands for a feature of a classified instance, and each branch denotes the value that can be assumed by a node. The core of DT is a greedy algorithm, which builds trees in a top-down learning approach (Quinlan, 1996). This algorithm classifies an instance from the root node and sorts it according to the feature values. The steps as follows give a brief introduction on the algorithm of the DT: (a) the algorithm first selects the most apposite attribute for the root node; (b) the second step is that the algorithm divides the instances into several subgroups, in which the instances of each subgroup need to have the identical attribute value; (c) Lastly, the steps are repeated on the subgroups individually again and again until all instances have the identical category (Saxena, 2017) (Fig. 1).

A number of areas have utilized the DT technique, including medical science, education, etc. For instance, Pandiangan et al. (2020) believe that the study period is crucial to evaluate a university's quality. To take appropriate actions to enhance the quality of the university, they use the DT and Naive Bayes algorithm to make the prediction and find that DT achieves the highest accuracy. DT algorithm is also widely used in the medical area, for example, the symptom classification of the Lupus disease (Gomathi & Narayani, 2015), coronary artery disease (Abdar et al., 2019) and covid (Rochmawati et al., 2020).

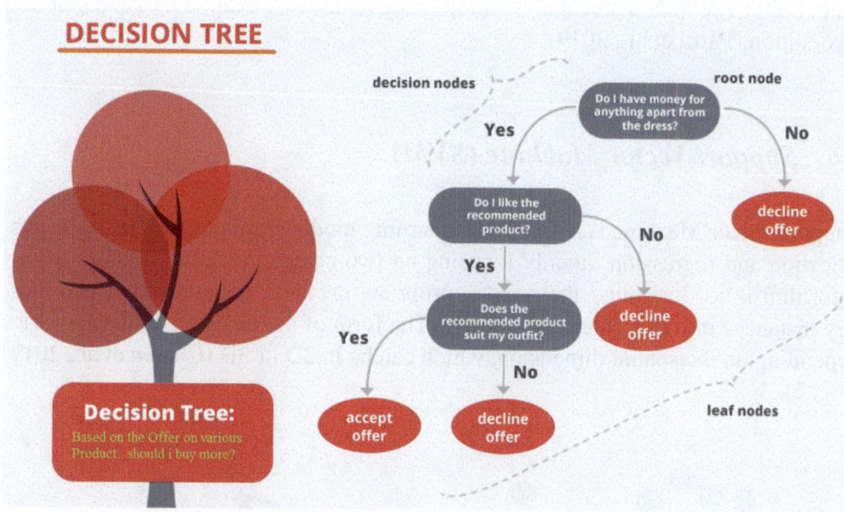

Fig. 1 Decision Tree Example

2.5 XGBoost Algorithm

Since its release in 2014, Extreme Gradient Boosting has been well received in the field of machine learning. Gradient Boosting Machines (GBMs) and XGBoost are both ensemble tree methods. XGBoost is short for Extreme Gradient Boosting. XGBoost is one of the decision tree algorithms, and it is used to deal with the issues of classification and regression. This method generates decision trees, with each successive tree attempting to decrease the errors of the preceding one. As a result, each tree constantly learns from the errors of its preceding tree and adjusts the residual errors. XGBoost combines computer code and hardware optimization approaches suitably to improve performance with less process power and time. Nevertheless, XGBoost is established on the foundation of GBM architecture through system improvements and recursive augmentations (Ohri, 2021).

Recently, a number of researchers in different areas have moved their attention to the application of XGBoost. Akter and Islam (2021) employed XGBoost to classify dementia for the identification of Alzheimer's disease. The algorithm performed well with an accuracy rate of 81%, and it identified that "ageAtEntry" was the most important feature. XGBoost is also employed to deal with the cybersecurity issue. Gawali et al. (2020) used XGBoost and Hidden Markov Model (HMM) for intrusion detection to build classifiers of attack signatures. Their results indicated that XGBoost performs better than HMM. Numerous other areas have evaluated the effectiveness of XGBoost algorithms, such as the prediction of student performance (Asselman et al., 2021), prediction and analysis of housing rent (Ming et al., 2020) and emotion recognition (Parui et al., 2019).

2.6 Support Vector Machine (SVM)

Support Vector Machine is a supervised learning model employed to deal with classification and regression, mainly focusing on two-category issues. The aim of this algorithm is to determine the most appropriate decision boundary that can classify instances into two categories clearly. The form of the decision boundary varies depending on the spatial dimension, which can be in 2D or 3D (Charan et al., 2017) (Fig. 2).

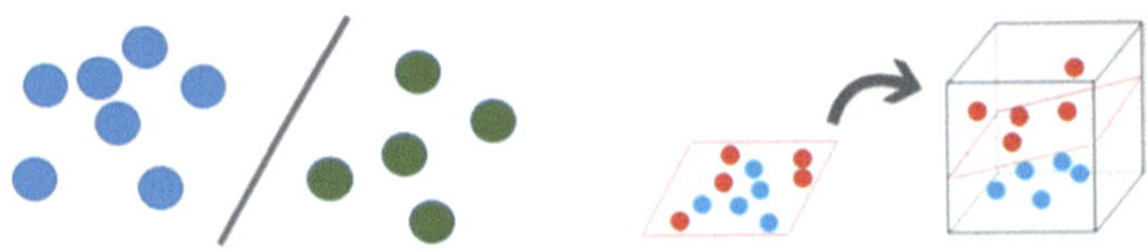

Fig. 2 Hyperplane

The most appropriate decision boundary should make the intervals between the boundary and the two groups of data, respectively, as large as possible. Support vectors are the data points closest to the boundary, and they determine the maximum interval. In addition, they play a vital role in the dataset as their position determines the position of the boundary. During the process of identifying the boundary, several boundary lines will be identified by only the one with the smallest bias is the decision boundary required. The SVM has also been employed in a variety of areas, for example, in marketing and sales (Syam and Kaul, 2021), in the prediction of student GPA (Dewi and Widiastuti, 2020), wind turbine event detection (Hu and Albertani, 2021).

3 Exploring Data and Choosing Features

The dataset is taken from the Kaggle Database, which is openly accessible and has been used by different authors because it includes sufficient observations. The dataset contains ten variables and 15,001 observations, and they are used to predict the presence or absence of diabetes in the patient. The variables are described in Table 1. Few of the tasks performed are as follows. First of all, histograms are plotted for data visualization. The histogram helps us to find the distribution of each feature in the dataset. The data is grouped in bins, and we can mention the number of bins we plan to visualize. The X-axis represents the range of the data values, and Y-axis represents the frequency.

The distplot code in the seaborn library is used to visualize for univariant analysis through a histogram, and it is a method of assessing the cumulative distribution function of a given variable (Figs. 3 and 4).

Table 1 The description of the data attributes

Attributes	Description
Patient Id	The identity of the user
Pregnancies	Whether a person gets diabetes after pregnancy
Plasma Glucose	The amount of glucose in the blood
DiastolicBloodPressure	The pressure when the heart is at rest
Triceps Thickness	The data of the body's fat storage
Serum Insulin	The amount of insulin in the body
BMI	Body Mass Index
Diabetes Pedigree	To know if anyone in the family has diabetes
Age	The age of the person
Diabetes	Whether the person has diabetes or not

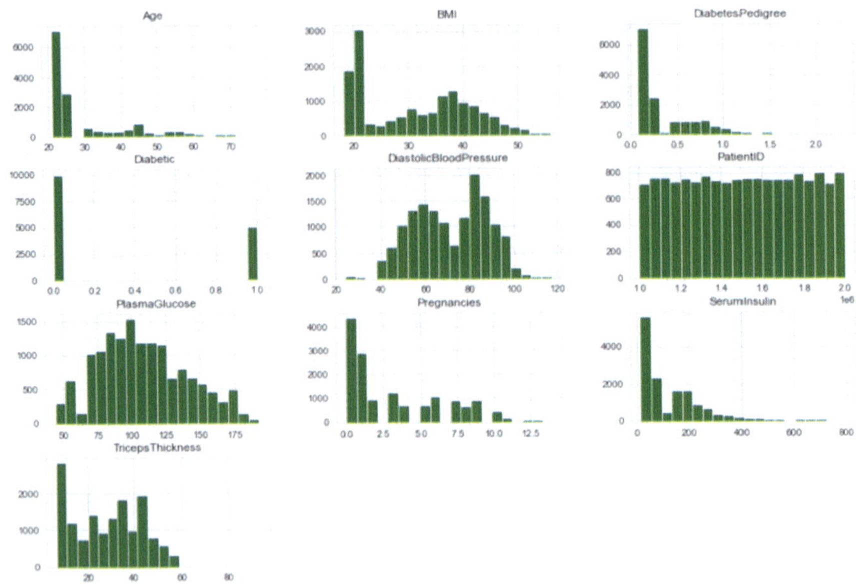

Fig. 3 Data Visualization

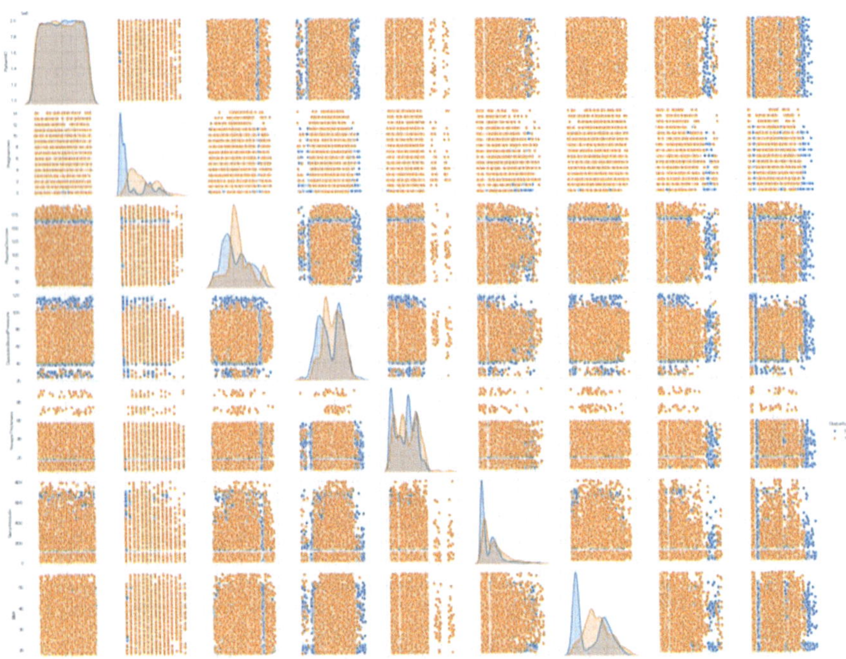

Fig. 4 The relation between the different columns in our dataset using pair plot

Afterward, the correlation analysis is used for finding the relation between the variables (features) using the heatmap. To be specific, Pearson Correlation is employed to conduct this analysis. The coefficient value of Pearson Correlation could have been between –1 and + 1. 1 suggests that they too are closely linked and 0 represents that there really is no correlation. In general, it is considered that the absolute value of correlation coefficients less than 0.3 indicates no correlation, the absolute value of correlation coefficients between 0.3 and 0.5 implies insignificant correlations, the absolute value of correlation coefficients between 0.5 and 0.8 indicates significant correlations and higher than 0.8 implies highly significant correlations. From the heatmap (Fig. 5), we can visualize no correlated variables as the highest value of correlation coefficients is 0.41, meaning that there is no multicollinearity issue in our dataset. We also check for null or duplicate values to see if the dataset needs to be cleaned further.

In addition, the Describe function is employed to obtain a general understanding of the data. This method works only with numeric values. Using this method, we get information of descriptive statistics like count, mean, standard deviation of the particular feature in the dataset, while min and max give us the minimum and maximum values. From Fig. 6, we can clearly see that there is no minimum value of zero(0). Moreover, 25, 50, and 75% are the quartiles of each feature.

When applying a particular algorithm to the data set, we apply the Function Scaling. It is necessary to scale the variables for distance-based models, so SVC is required to operate on distance metrics to carry out standardization.

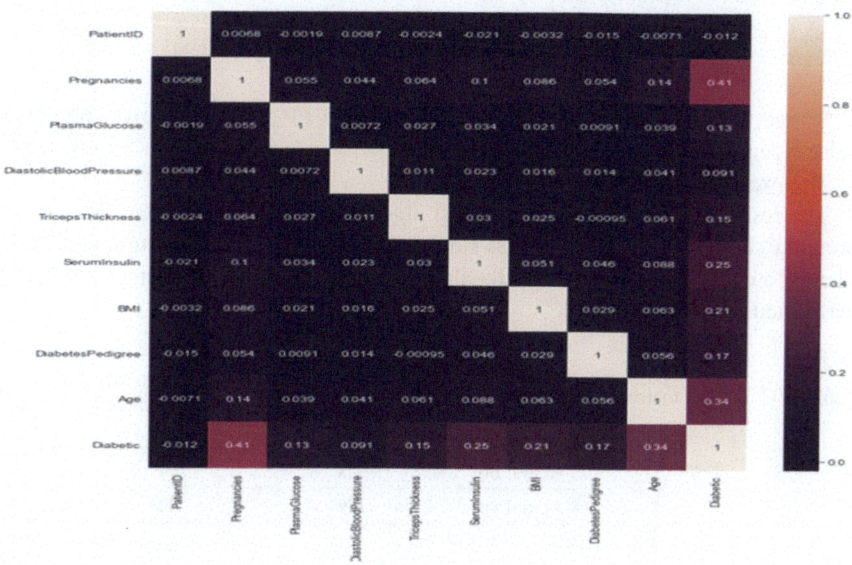

Fig. 5 For the correlation between the variables, we use a heatmap

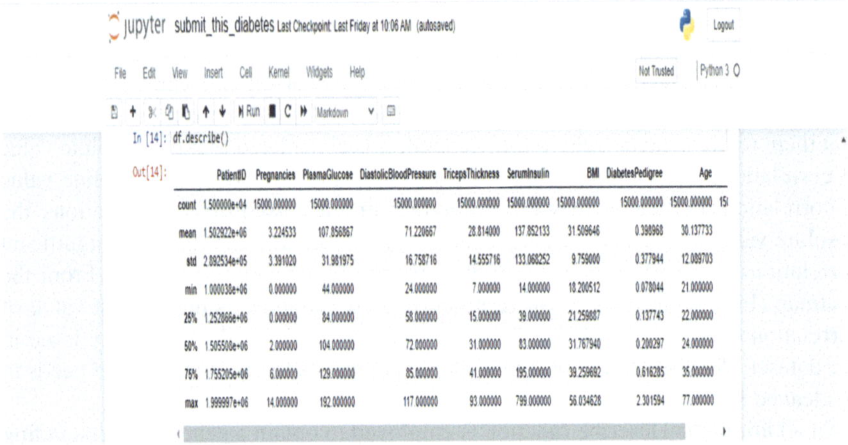

	PatientID	Pregnancies	PlasmaGlucose	DiastolicBloodPressure	TricepsThickness	SerumInsulin	BMI	DiabetesPedigree	Age
count	1.500000e+04	15000.000000	15000.000000	15000.000000	15000.000000	15000.000000	15000.000000	15000.000000	15000.000000
mean	1.502922e+06	3.224533	107.856867	71.220667	28.814000	137.852133	31.509646	0.398968	30.137733
std	2.892534e+05	3.391020	31.981975	16.758716	14.555716	133.068252	9.759000	0.377944	12.089703
min	1.000038e+06	0.000000	44.000000	24.000000	7.000000	14.000000	18.200512	0.078044	21.000000
25%	1.252866e+06	0.000000	84.000000	58.000000	15.000000	39.000000	21.259887	0.137743	22.000000
50%	1.505508e+06	2.000000	104.000000	72.000000	31.000000	83.000000	31.767940	0.200297	24.000000
75%	1.755205e+06	6.000000	129.000000	85.000000	41.000000	195.000000	39.259692	0.616285	35.000000
max	1.999997e+06	14.000000	192.000000	117.000000	93.000000	799.000000	56.034628	2.301594	77.000000

Fig. 6 Using describe a method for the dataset

4 Experiment and Results

The dataset is divided into the training dataset and testing dataset with a ratio of 70:30. We use four machine learning algorithms for this dataset, including Logistic Regression, SVC, XGBoost and Decision Tree. Moreover, the accuracy can be calculated using the below formula:

$$\text{Accuracy} = (TP + TN)/(TP + TN + FP + FN).$$

Where, TP = True Positive, TN = True Negative, FP = False Positive and FN = False Negative.

The example is as below:

The results showed that the XGBoost algorithm best fixed this dataset with an accuracy of 95%, and the worst was Logistic Regression with an accuracy of 79%. The following part explains how the algorithms are used and what steps are being performed on them (Table 2).

Table 2 Confusion matrix

	Predicted no	Predicted yes
Actual no	TN	FP
Actual yes	FN	TP

Table 3 Confusion matrix of logistic regression

	Predicted no	Predicted yes
Actual no	2966	365
Actual yes	642	977

4.1 Logistic Regression

The linear and logistic algorithms are similar but differ in functionalities. Logistic Regression explores and describes the relationship between the group of dependent attributes and one or more attributes (Hackernoon, 2020).

This technique of algorithms is widely used in dataset analysis and prediction. The parameter depending on is always a binary variable in this algorithm. This is used mainly for estimation and for measuring overall performance. Logistic Regression has less accuracy when compared other two algorithms.

The Confusion Matrix of Logistic Regression is as follows in Table 3, and the accuracy can be computed, which is 79.65%.

4.2 Decision Tree Classifier

The Decision Tree algorithm is part of the community of supervised learning. The level of comprehension of the Decision Trees algorithm is limited compared to other classification models. The decision tree algorithm attempts to fix the problem using a tree structure (Medium, 2020). The actual root of the tree belongs to the category, but every leaf node matches the sign of the community. The Decision Tree Classifier seems to be the second method with sufficient variance.

This project also applies Decision Tree Classifier to the data of diabetes. Its Confusion Matrix is presented in Table 4, and the accuracy is 90.54%, which is higher than Logistic Regression.

Table 4 Confusion matrix of decision Tree

	Predicted no	Predicted yes
Actual no	3095	236
Actual yes	232	1387

Table 5 Confusion matrix of XGBoost

	Predicted no	Predicted yes
Actual no	3220	111
Actual yes	123	1496

Table 6 Confusion matrix of SVC

	Predicted no	Predicted yes
Actual no	3096	235
Actual yes	337	1282

4.3 XGBoost Algorithm

The purpose of the XGBoost was to make efficient use of the resources available to construct the model (Reinstein, 2020). There are two key reasons for using XGBoost with the experiment, including fast execution and outstanding performance of the model. It is an extremely powerful and dynamic method that can function through several challenges of regression, classification, rating and fitness function developed by the clients. This XGBoost algorithm also does not require the scaling process. Moreover, this algorithm is known to have higher accuracy relative to the other three algorithms used. The Confusion Matrix of XGBoost is as follows in Table 5, and the accuracy is 95.27%.

4.4 SVC (Support Vector Classifier)

SVC (Support Vector Classifier) is a designed study method used in data analysis for regression and classification approaches. It can actually fix linear or nonlinear concerns and operates quite well for complex applications. It aims to match the target with the information you supply and return a "best suitable" hyper-plane separating or classifying your data.

With this, you can then collect a few attributes of your categorization when you get a hyper-plane to see the "expected" level. The hyperplane is determined based on the maximum range from the exercise set of data for a greater variety, which decreases the error of the sentences. The Confusion Matrix of SVC is as follows in Table 5, and its accuracy is 88.44% (Table 6).

Table 7 The summary of the algorithms' accuracies

	Logistic regression	Decision tree	XGBoost	SVC
Accuracy	79.65	90.54	95.27	88.44

5 Discussions

In the above section, this project discusses the machine learning algorithms used to analyze the data of diabetes. The summary of their accuracy is shown in Table 7.

Depending on the outcome, the XGBoost algorithm received the highest accuracy of 95.27 percent based on the target term, the algorithm which contains the presence or absence of diabetes. The second-best algorithm in this report is the Decision Tree Classifier, with an accuracy of 90.54 percent. These four algorithms efficiently considered that the most pressing issue is to determine diabetes disease based on the chosen features that are responsive. In addition, it is important to identify or improve the machine learning algorithms that can acquire nearly 100% accuracy that eventually saves many people's lives. In this report, lesser attention is given to Logistic Regression.

6 Conclusion

In this paper, diabetes machine learning algorithms were employed to analyze the dataset DAT263x Lab01 that has been collected from the Kaggle database. The data comprises observations of 15,001 and 10 variables without null values or duplicated values. The algorithms that have been used are Logistic Regression, SVC, XGBoost, and Decision Tree Classifier. Before conducting the ML algorithms, several tasks were performed to obtain an understanding of the dataset used, such as data visualization, correlation analysis and descriptive analysis.

Confusion matrixes were extracted and explained by dividing the data into the training and testing datasets. The results obtained from the algorithms chosen prove that Xgboost and Decision Tree Classifier are the best algorithms with 95 and 90% accuracy, respectively.

The application of data mining provides a facility for early diagnosis of diabetes and potentially reduces the loss of life as a result, especially during health care events like the New Coronavirus pandemic. SARS-CoV-2 is a threat to diabetic patients. Using these technologies can protect lives in the event of an epidemic, such as during the new crown epidemic. A quick determination of a patient's condition can prevent complications to a certain extent and even avoid death. In addition, due to the increase in COVID-19 patients, medical institutions are under a lot of pressure, leading to disruptions in the treatment and diagnosis of other diseases as well. Therefore, our prediction model is able to make the early diagnosis of diabetes more effective and efficient.

Acknowledgements This work is partly supported by VC Research (VCR 0000156).

Appendix

The program is composed in a python programed on a Jupiter notebook. That necessary resources are mentioned throughout the code as provided by the algorithm.

Logistic Regression:

```
#importing all the neccessary libararies

import pandas as pd
import numpy as np
import matplotlib.pyplot as plt
import seaborn as sns

#loading the dataset and reading using pandas
df=pd.read_csv('C:/Users/Saiteja/Downloads/diabetes-from-dat263x-lab01/diabetes.csv')

#dispalying the first 10 values in the dataset
df.head(10)

#dtypes gives the datatype of the data
df.dtypes

#Data Visualization
df.hist(figsize=(15,10),bins=20,color='green')

#to understand the relation between the different columns (features ) in our dataset we use
pair plot (hue as outcome)
sns.set_style('whitegrid')
sns.pairplot(df,hue='Diabetic')
```

```
#to find the correlation between the variables we use heatmap
fig = plt.figure(figsize = (10,10))
sns.heatmap(df.corr(), annot = True)


#plotting the distplots of the features

fig = plt.figure(figsize=(15,10))
fig.add_subplot(231)
plt.title('SerumInsulin', fontsize=15)
sns.distplot(df['SerumInsulin'], bins = 20, kde=True)
fig.add_subplot(232)
plt.title('BMI', fontsize=15)
sns.distplot(df['BMI'], bins = 20, kde=True)
fig.add_subplot(233)
plt.title('Age', fontsize=15)
sns.distplot(df['Age'], bins = 20, kde=True)
fig.add_subplot(234)
plt.title('Pregnancies', fontsize=15)
sns.distplot(df['Pregnancies'], bins = 20, kde=True)
fig.add_subplot(235)
plt.title('PlasmaGlucose', fontsize=15)
sns.distplot(df['PlasmaGlucose'], bins = 20, kde=True)
fig.add_subplot(236)
plt.title('DiastolicBloodPressure', fontsize=15)
sns.distplot(df['DiastolicBloodPressure'], bins = 20, kde=True)
fig = plt.figure(figsize=(10,10))
fig.add_subplot(221)
plt.title('DiabetesPedigree', fontsize=15)
sns.distplot(df['DiabetesPedigree'], bins = 20, kde=True)

fig.add_subplot(222)
plt.title('TricepsThickness', fontsize=15)
sns.distplot(df['TricepsThickness'], bins = 20, kde=True)

#gives the count here we are visualizing the count of the target variable
```

```
sns.countplot(x = 'Diabetic', data = df)

sns.heatmap(df.isnull(), yticklabels=False)
```

#checking whether the dataset has duplicates or not

```
df.duplicated().sum()
```

#check for null values in the dataset

```
df.isnull().sum()

df.describe()

df.describe()
```

 # Segregating the input data and the output parameter

```
#X contains all the independent variables(input variables)excluding the PatientID as it is not required to train the model
#input features are used to predict the target varaibale ('Diabetic')
#Y contain the target variable(Dependent varibale('Diabetic'))
X = df.iloc[:,1:9]
y = df.iloc[:,9]
```

#checking X(input variables after seperating)

```
X
```

#target variable which is sepearted from the input varibales

```
Y
```

#Splitting the dataset into the Training set and Test set

```
from sklearn.model_selection import train_test_split
X_train, X_test, y_train, y_test = train_test_split(X, y, test_size=0.33, random_state=56)
```

Applying Logistic

```
From sklearn.linear_model import LogisticRegression
logmodel = LogisticRegression()
logmodel.fit(X_train,y_train)
y_log_pred = logmodel.predict(X_test)
```

#importing all the neccessary libararies required for accuracy

```
from sklearn.metrics import confusion_matrix
from sklearn.metrics import f1_score
from sklearn.metrics import accuracy_score
from sklearn.metrics import classification_report
```

#accuracy metrics

```
print(confusion_matrix(y_test,y_log_pred))
print(classification_report(y_test,y_log_pred))
log_accuracy = accuracy_score(y_log_pred, y_test)
print(log_accuracy*100)
```

#visualizing the confusion matrix for logistic

```
cm = confusion_matrix(y_test,y_log_pred)
sns.heatmap(cm, annot = True, fmt='g')
```

Decision Tree Classifier

#importing all the neccessary libararies

```
import pandas as pd
import numpy as np
import matplotlib.pyplot as plt
import seaborn as sns
```

#loading the dataset and reading using pandas

```
df=pd.read_csv('C:/Users/Saiteja/Downloads/diabetes-from-dat263x-lab01/diabetes.csv')
```

#dispalying the first 10 values in the dataset

```
df.head(10)
```

#dtypes gives the datatype of the data

```
df.dtypes
```

#Data Visualization

```
df.hist(figsize=(15,10),bins=20,color='green')
```

#to understand the relation between the different columns (features) in our dataset we use pair plot (hue as outcome)

```
sns.set_style('whitegrid')
sns.pairplot(df,hue='Diabetic')
```

#to find the correlation between the variables we use heatmap

```
fig = plt.figure(figsize = (10,10))
sns.heatmap(df.corr(), annot = True)
```

#plotting the distplots of the features

```
fig = plt.figure(figsize=(15,10))
fig.add_subplot(231)
plt.title('SerumInsulin', fontsize=15)
sns.distplot(df['SerumInsulin'], bins = 20, kde=True)
fig.add_subplot(232)
plt.title('BMI', fontsize=15)
sns.distplot(df['BMI'], bins = 20, kde=True)
fig.add_subplot(233)
plt.title('Age', fontsize=15)
sns.distplot(df['Age'], bins = 20, kde=True)
fig.add_subplot(234)
plt.title('Pregnancies', fontsize=15)
sns.distplot(df['Pregnancies'], bins = 20, kde=True)
fig.add_subplot(235)
plt.title('PlasmaGlucose', fontsize=15)
sns.distplot(df['PlasmaGlucose'], bins = 20, kde=True)
fig.add_subplot(236)
plt.title('DiastolicBloodPressure', fontsize=15)
sns.distplot(df['DiastolicBloodPressure'], bins = 20, kde=True)

fig = plt.figure(figsize=(10,10))
fig.add_subplot(221)
plt.title('DiabetesPedigree', fontsize=15)
sns.distplot(df['DiabetesPedigree'], bins = 20, kde=True)
fig.add_subplot(222)
plt.title('TricepsThickness', fontsize=15)
sns.distplot(df['TricepsThickness'], bins = 20, kde=True)
```

```
#gives the count here we are vizualising the count of the target variable
sns.countplot(x = 'Diabetic', data = df)

sns.heatmap(df.isnull(), yticklabels=False)

#checking whether the dataset has duplicates or not
df.duplicated().sum()

#check for null values in the dataset
df.isnull().sum()

df.describe()

df.describe()

#seperating input variables and the target variable

#X contains all the independent variables(input variables)excluding the PatientID as it is not
required to train the model
#input features are used to predict the target varaibale ('Diabetic')
#Y contain the target variable(Dependent varibale('Diabetic'))
X = df.iloc[:,1:9]
y = df.iloc[:,9]

#checking X(input variables after seperating)
X

#target variable which is sepearted from the input varibales

Y
```

#Applying Decision Tree algorithm

```
from sklearn.tree import DecisionTreeClassifier
#Creating an instance of DecisionTreeClassifier() called dec_tree
dec_tree = DecisionTreeClassifier()
```

#fitting the model to the training data

```
dec_tree.fit(X_train,y_train)
```

#predictions from the test data

```
y_dec_pred = dec_tree.predict(X_test)

print(confusion_matrix(y_test,y_dec_pred))
print(classification_report(y_test,y_dec_pred))
dec_tree_accuracy = accuracy_score(y_dec_pred, y_test)
print(dec_tree_accuracy*100)

cm = confusion_matrix(y_test,y_dec_pred)
sns.heatmap(cm, annot = True, fmt='g'
```

Xgboost Classifier

#importing all the neccessary libararies

```
import pandas as pd
import numpy as np
import matplotlib.pyplot as plt
import seaborn as sns
```

#loading the dataset and reading using pandas

```
df=pd.read_csv('C:/Users/Saiteja/Downloads/diabetes-from-dat263x-lab01/diabetes.csv')
```

```
#dispalying the first 10 values in the dataset
df.head(10)

#dtypes gives the datatype of the data
df.dtypes

#Data Visualization
df.hist(figsize=(15,10),bins=20,color='green')

#to understand the relation between the different columns (features ) in our dataset we use
pair plot (hue as outcome)
sns.set_style('whitegrid')
sns.pairplot(df,hue='Diabetic')

#to find the correlation between the variables we use heatmap
fig = plt.figure(figsize = (10,10))
sns.heatmap(df.corr(), annot = True)

#plotting the distplots of the features

fig = plt.figure(figsize=(15,10))
fig.add_subplot(231)
plt.title('SerumInsulin', fontsize=15)
sns.distplot(df['SerumInsulin'], bins = 20, kde=True)
fig.add_subplot(232)
plt.title('BMI', fontsize=15)
sns.distplot(df['BMI'], bins = 20, kde=True)
fig.add_subplot(233)
plt.title('Age', fontsize=15)
sns.distplot(df['Age'], bins = 20, kde=True)
fig.add_subplot(234)
plt.title('Pregnancies', fontsize=15)
sns.distplot(df['Pregnancies'], bins = 20, kde=True)
fig.add_subplot(235)
plt.title('PlasmaGlucose', fontsize=15)
sns.distplot(df['PlasmaGlucose'], bins = 20, kde=True)
fig.add_subplot(236)
plt.title('DiastolicBloodPressure', fontsize=15)
sns.distplot(df['DiastolicBloodPressure'], bins = 20, kde=True)
```

```
fig = plt.figure(figsize=(10,10))
fig.add_subplot(221)
plt.title('DiabetesPedigree', fontsize=15)
sns.distplot(df['DiabetesPedigree'], bins = 20, kde=True)


fig.add_subplot(222)
plt.title('TricepsThickness', fontsize=15)
sns.distplot(df['TricepsThickness'], bins = 20, kde=True)
```

#gives the count here we are vizualising the count of the target variable

```
sns.countplot(x = 'Diabetic', data = df)

sns.heatmap(df.isnull(), yticklabels=False)
```

#checking whether the dataset has duplicates or not

```
df.duplicated().sum()
```

#check for null values in the dataset

```
df.isnull().sum()

df.describe()

df.describe()
```

#seperating input variables and the target variable

```
#X contains all the independent variables(input variables)excluding the PatientID as it is not required to train the model
#input features are used to predict the target varaibale ('Diabetic')
#Y contain the target variable(Dependent varibale('Diabetic'))
X = df.iloc[:,1:9]
y = df.iloc[:,9]
```

#checking X(input variables after seperating)

```
X
```

#target variable which is sepearted from the input varibales

```
Y
```

#Applying Xgboost

 # Fitting XGBoost to the Training set and predicting the results

```
import xgboost
classifier = xgboost.XGBClassifier()
classifier.fit(X_train, y_train)

y_xg_pred = classifier.predict(X_test)
print(confusion_matrix(y_test, y_xg_pred))
print(classification_report(y_test,y_xg_pred))
print(accuracy_score(y_test,y_xg_pred)*100)

cm = confusion_matrix(y_test,y_xg_pred)
sns.heatmap(cm, annot = True, fmt='g')
```

Support Vector Classifier

#importing all the neccessary libararies

```
import pandas as pd
import numpy as np
import matplotlib.pyplot as plt
import seaborn as sns
```

#loading the dataset and reading using pandas

```
df=pd.read_csv('C:/Users/Saiteja/Downloads/diabetes-from-dat263x-lab01/diabetes.csv')
```

#dispalying the first 10 values in the dataset

```
df.head(10)
```

#dtypes gives the datatype of the data

```
df.dtypes
```

#Data Visualization

```
df.hist(figsize=(15,10),bins=20,color='green')
```

```
#to understand the relation between the different columns (features ) in our dataset we use
pair plot (hue as outcome)
sns.set_style('whitegrid')
sns.pairplot(df,hue='Diabetic')

#to find the correlation between the variables we use heatmap
fig = plt.figure(figsize = (10,10))
sns.heatmap(df.corr(), annot = True)

#plotting the distplots of the features

fig = plt.figure(figsize=(15,10))
fig.add_subplot(231)
plt.title('SerumInsulin', fontsize=15)
sns.distplot(df['SerumInsulin'], bins = 20, kde=True)
fig.add_subplot(232)
plt.title('BMI', fontsize=15)
sns.distplot(df['BMI'], bins = 20, kde=True)
fig.add_subplot(233)
plt.title('Age', fontsize=15)
sns.distplot(df['Age'], bins = 20, kde=True)
fig.add_subplot(234)
plt.title('Pregnancies', fontsize=15)
sns.distplot(df['Pregnancies'], bins = 20, kde=True)
fig.add_subplot(235)
plt.title('PlasmaGlucose', fontsize=15)
sns.distplot(df['PlasmaGlucose'], bins = 20, kde=True)
fig.add_subplot(236)
plt.title('DiastolicBloodPressure', fontsize=15)
sns.distplot(df['DiastolicBloodPressure'], bins = 20, kde=True)

fig = plt.figure(figsize=(10,10))
fig.add_subplot(221)
plt.title('DiabetesPedigree', fontsize=15)
sns.distplot(df['DiabetesPedigree'], bins = 20, kde=True)


fig.add_subplot(222)
plt.title('TricepsThickness', fontsize=15)
sns.distplot(df['TricepsThickness'], bins = 20, kde=True)
```

```
#gives the count here we are vizualising the count of the target variable
sns.countplot(x = 'Diabetic', data = df)

sns.heatmap(df.isnull(), yticklabels=False)

#checking whether the dataset has duplicates or not
df.duplicated().sum()

#check for null values in the dataset
df.isnull().sum()

df.describe()

df.describe()

#seperating input variables and the target variable

#X contains all the independent variables(input variables)excluding the PatientID as it is not required to train the model
#input features are used to predict the target varaibale ('Diabetic')
#Y contain the target variable(Dependent varibale('Diabetic'))
X = df.iloc[:,1:9]
y = df.iloc[:,9]

#checking X(input variables after seperating)
X

#target variable which is sepearted from the input varibales

Y
```

Applying support vector classifier

```
from sklearn.preprocessing import StandardScaler

sc = StandardScaler()
```

```
X_train = sc.fit_transform(X_train)

X_test = sc.transform(X_test)
```

#applying support vector classifier

```
from sklearn.svm import SVC

s_model = SVC()
s_model.fit(X_train,y_train)

svm_predicted = s_model.predict(X_test)


print(confusion_matrix(y_test,svm_predicted))

print(classification_report(y_test,svm_predicted))

svm_accuracy = accuracy_score(svm_predicted, y_test)

print(svm_accuracy*100)


cm = confusion_matrix(y_test,svm_predicted)

sns.heatmap(cm, annot = True, fmt='g')
```

References

Abdar, M., Nasarian, E., Zhou, X., Bargshady, G., Wijayaningrum, V. N., & Hussain, S. (2019). Performance improvement of decision trees for diagnosis of coronary artery disease using multi filtering approach. In *2019 IEEE 4th International Conference on Computer and Communication Systems (ICCCS)* (pp. 26–30). Singapore. https://doi.org/10.1109/CCOMS.2019.8821633

Akter, L., & Ferdib-Al-Islam. (2021). Dementia identification for diagnosing Alzheimer's disease using XGBoost algorithm. In *2021 International Conference on Information and Communication Technology for Sustainable Development (ICICT4SD)* (pp. 205–209).

American Diabetes Association. (2021). How COVID-19 Impacts People with Diabetes. Available online: https://www.diabetes.org/coronavirus-covid-19/how-coronavirus-impacts-people-with-diabetes. Retrieved January 3, 2021.

Asselman, A., Khaldi, M., & Aammou, S. (2021). Enhancing the prediction of student performance based on the machine learning XGBoost algorithm. *Interactive Learning Environments,* 1–20

Boyd, C. R., Tolson, M. A., & Copes, W. S. (1987). Evaluating trauma care: The TRISS method. Trauma score and the injury severity score. *The Journal of Trauma., 27*(4), 370–378. https://doi.org/I0.1097/00005373-198704000-00005.PMID3106646

Charan, R., Manisha. A., Ravichandran, K., & Muthu, R. (2017). A text-independent speaker verification model: A comparative analysis. In *2017 IEEE International Conference on Intelligent Computing and Control (I2C2)*. India. https://doi.org/10.1109/I2C2.2017.8321794.

Chaves, L., & Marques, G. (2021). Data mining techniques for early diagnosis of diabetes: A comparative study. *Applied Sciences, 11*(5), 2218.

Chitra, R., & Seenivasagam, V. (2013). Review of heart disease prediction system using data mining and hybrid intelligent techniques. *ICTACT Journal on Soft Computing, 3*(04), 605–609.

Christodoulou, E., Ma, J., Collins, G. S., Steyerberg, E. W., Verbakel, J. Y., & Calster, B. V. (2019). A systematic review shows no performance benefit of machine learning over logistic regression for clinical prediction models. *Journal of Clinical Epidemiology, 110*, 12–22.

Deng, M., Jiang, L., Ren, Y., & Liao, J. (2020). Can we reduce mortality of COVID-19 if we do better in glucose control? *Medicine in Drug Discovery, 7*(2020), 100048.

Dewi, K. E., & Widiastuti, N. I. (2020, July). Support vector regression for GPA prediction. In *IOP Conference Series: Materials Science and Engineering* (Vol. 879, No. 1, p. 012112). IOP Publishing.

Fitriyani, N., Syafrudin, M., Alfian, G., & Rhee, J. (2019). Development of disease prediction model based on ensemble learning approach for diabetes and hypertension. *IEEE Access., 7*, 144777–144789. https://doi.org/10.1109/ACCESS.2019.2945129

Gawali, S., Agale, P., Ghorpade, S., Gawade, R., Nimat, P. (2020). Intrusion detection using hidden Markov model and XGBoost algorithm. *International Journal of Scientific Research in Computer Science, Engineering and Information Technology*, 466–470. https://doi.org/10.32628/CSEIT206287.

Gomathi, S., & Narayani, V. (2015). Monitoring of Lupus disease using decision tree induction classification algorithm. In *2015 International Conference on Advanced Computing and Communication Systems* (pp. 1–6). Coimbatore, India. https://doi.org/10.1109/ICACCS.2015.7324054

Hackernoon.com. (2020). Introduction to Machine Learning Algorithms: Logistic *Regressio*|Hacker Noon. [online] Available at: https://hackernoon.com/introduction-to-machine-learning-algorithms-logistic-regression-cbdd82d81a36. Retrieved August 10, 2020.

Hartmann-Boyce, J., Morris, E., Goyder, C., Kinton, J., Perring, J., Nunan, D., & Khunti, K. (2020). Diabetes and COVID-19: risks, management, and learnings from other national disasters. *Diabetes Care, 43*(8), 1695–1703

Hu, C., & Albertani, R. (2021). Wind turbine event detection by support vector machine. *Wind Energy, 24*(7), 672–685.

Kologlu M., Elker D., Altun H., & Sayek I. (2001). Validation of MPI and OIA II in two different groups of patients with secondary peritonitis II. *Hepato-Gastroenterology, 48*, N2 37. 147–151.

Komi, M., Li, J., Zhai, Y., & Zhang, X. (2017). Application of data mining methods in diabetes prediction. In *Presented at the 2017 2nd International Conference on Image, Vision and Computing (ICIVC)* (pp. 1006–1010). Chengdu, China: IEEE. https://doi.org/10.1109/ICIVC.2017.7984706

Kumar, P. S., Kumari, A., Mohapatra, S., Naik, B., Nayak, J., & Mishra, M. (2021). CatBoost ensemble approach for diabetes risk prediction at early stages. In *Presented at the 2021 1st Odisha International Conference on Electrical Power Engineering, Communication and Computing Technology(ODICON)* (pp. 1–6). Bhubaneswar, India: IEEE. https://doi.org/10.1109/ODICON50556.2021.9428943

Lim, S., Bae, J. H., Kwon, H. S., & Nauck, M. A. (2021). COVID-19 and diabetes mellitus: From pathophysiology to clinical management. *Nature Reviews Endocrinology, 17*(1), 11–30.

Medicalnewstoday.com. 2020. Diabetes: Symptoms, Treatment, And Early Diagnosis. [online] Available at: https://www.medicalnewstoday.com/articles/323627. Retrieved August 10, 2020.

Medium. (2020). Decision Tree Algorithm — Explained. [online] Available at: https://towardsdatascience.com/decision-tree-algorithm-explained-83beb6e78ef4. Retrieved August 12, 2020.

Ming, Y., Zhang, J., Qi, J., Liao, T., Wang, M., & Zhang, L. (2020, September). Prediction and analysis of Chengdu housing rent based on XGBoost algorithm. In *Proceedings of the 2020 3rd International Conference on Big Data Technologies* (pp. 1–5).

nhs.uk. (2020). Diabetes. [online] Available at: https://www.nhs.uk/conditions/diabetes/. Retrieved August 10, 2020.

NIDDK. (2020). *What Is Diabetes?|NIDDK*. [online] National Institute of Diabetes and Digestive and Kidney Diseases. Available at: https://www.niddk.nih.gov/health-information/diabetes/overview/what-is-diabetes. Retrieved August 10, 2020.

Ohri, A. (2021). XGBoost Algorithm: An Easy Overview For 2021. Available at: XGBoost Algorithm: An Easy Overview For 2021 (jigsawacademy.com). Retrieved June 15.

Pandiangan, N., Buono, M. L. C., & Loppies, S. H. D. (2020). Implementation of decision tree and Naïve Bayes classification method for predicting study period. *Journal of Physics: Conference, 1569*, 022022. https://doi.org/10.1088/1742-6596/1569/2/022022

Parui, S., Bajiya, A. K. R., Samanta, D., & Chakravorty, N. (2019, December). Emotion recognition from EEG signal using XGBoost algorithm. In *2019 IEEE 16th India Council International Conference (INDICON)* (pp. 1–4). IEEE.

Quinlan, J. R. (1996). Learning decision tree classifiers. *ACM Computing Surveys (CSUR), 28*(1), 71–72.

Rashid, M., Singh, H., Goyal, V., Parah, S. A., & Wani, A. R. (2021). Big data based hybrid machine learning model for improving performance of medical Internet of Things data in healthcare systems. In *Healthcare Paradigms in the Internet of Things Ecosystem* (pp. 47–62). Academic Press.

Reinstein, I. (2020). Xgboost, A Top Machine Learning Method On Kaggle, Explained—Kdnuggets. [online] KDnuggets. Available at: https://www.kdnuggets.com/2017/10/xgboost-top-machine-learning-method-kaggle-explained.html. Retrieved August 10, 2020.

Rochmawati, N., Hidayati, H. B., Yamasari, Y., Yustanti, W., Rakhmawati, L., Tjahyaningtijas, H. P. A., & Anistyasari, Y. (2020). Covid Symptom Severity Using Decision Tree. In *2020 Third International Conference on Vocational Education and Electrical Engineering (ICVEE)* (pp. 1–5). Indonesia: Surabaya. https://doi.org/10.1109/ICVEE50212.2020.9243246

Samant, P., & Agarwal, R. (2018a). Machine learning techniques for medical diagnosis of diabetes using iris images. *Computer Methods and Programs in Biomedicine, 157*, 121–128. https://doi.org/10.1016/j.cmpb.2018.01.004

Samant, P., & Agarwal, R. (2018b). Comparative analysis of classification based algorithms for diabetes diagnosis using iris images. *Journal of Medical Engineering and Technology, 42*, 35–42. https://doi.org/10.1080/03091902.2017.1412521

Saxena, R. (2017). *How Decision Tree Algorithm Works*. Available at: https://dataaspirant.com/2017/01/30/how-decision-tree-algorithm-works/. Retrieved 04, April 40.

Sisodia, D., & Sisodia, D. S. (2018). Prediction of diabetes using classification algorithms. *Procedia Computer Science, 132*, 1578–1585.

Swapna, G., Soman, K. P., & Vinayakumar, R. (2018). Automated detection of diabetes using CNN and CNN-LSTM network and heart rate signals. *Procedia Computer Science, 132*, 1253–1262.

Syam, N., & Kaul, R. (2021). Support vector machines in marketing and sales. In *Machine learning and artificial intelligence in marketing and sales*. Emerald Publishing Limited.

WHO. Available at :https://covid19.who.int/. Retrieved October 27, 2021.

Wu, F., Zhao, S., Yu, B., Chen, Y. M., Wang, W., Song, Z. G., & Zhang, Y. Z. (2020). A new coronavirus associated with human respiratory disease in China. *Nature, 579*(7798), 265–269.

Yang, J. K., Lin, S. S., Ji, X. J., & Guo, L. M. (2010). Binding of SARS coronavirus to its receptor damages islets and causes acute diabetes. *Acta Diabetologica, 47*(3), 193–199.

Impact of Covid-19 on Indian Agriculture

Debabrata Mondal

Abstract India's agriculture and its allied activities are the largest sources of livelihood for about 70 to 80% of the population. The COVID-19 virus is spreading rapidly and India declared a 4th phase nation-wise lockdown until May 31. Due to this lockdown, many have lost their jobs and are unable to continue to earn their livelihood. Famers incurred losses due to non-availability of labor as well as lack of transportation facilities and market closure. Poultry farmers reported that they incurred losses due to a huge stock of unsold eggs and were unable to go to the market to sell their products. Fishermen were not able to fish and were very much worried about the fishing ban in seas and rivers, subsequently delayed harvesting. Market closure and movement restrictions affected their job and incurred a huge loss. It seems that they are facing live risk for not gaining employment. Ministry of Home Affairs, Government of India, announced that exempted agriculture and allied activities from lockdown restrictions. Laborers can go to work to generate their livelihood, and markets are open, Mahatma Gandhi National Rural Employment Guarantee Scheme (MGNREGS) work will also commence. However, there is some transportation/ communication gap at the village level. Mostly impacted livelihoods are those who do businesses or engage in casual labor or petty trade. Therefore, the government should have a focus on the lives of citizens of every section of society. This paper examined impacts on livelihoods, food security, access to goods and materials. It highlights how people are anxious about making daily needs and staying healthy amid unprecedented constraints. Though, people who generate their livelihood on agriculture and allied activities have lost their income and are mostly jobless. Many opted for other livelihood options such as daily wagers, vegetable sellers, and homemakers to sustain life and family. The shortage of laborers could disrupt the production and processing of food, business, industries, and other sectors. A total of 400 respondents were selected for this study. This paper analyzed the impact of COVID-19 on different sections of people. The use of Information

D. Mondal (✉)
Research Scholar, Department of Agricultural Extension, Institute of Agriculture, Visva Bharati, Sriniketan, West Bengal 731236, India
e-mail: mondal.debabrata@gmail.com

V. Chang et al. (eds.), *Artificial Intelligence and Machine Learning Methods in COVID-19 and Related Health Diseases*, Studies in Computational Intelligence 1023,
https://doi.org/10.1007/978-3-031-04597-4_9

and Communication Technologies (ICTs) can boost up agricultural productivity and supply chain.

Keywords COVID-19 · Agriculture · Livelihood · Lockdown · ICTs

1 Introduction

The government of India declared a three weeks nationwide lockdown starting from March 25, 2020, for its 1st phase to cope with the spread of the pandemic. All the services and businesses, educational institutions, religious places, public utilities, and government offices across the country will be closed during this period and all means of transportation/ communication facilities have been stopped. Most of the migrant workers have not returned to their native places and are staying at their workplaces. As a result, they lose their job and don't have enough money to continue their livelihood. So their life is at risk (Alankar et al., 2021; Dev & Sengupta, 2020a, b).

About 80% of the Indian farmers belong to small and marginal (up to 2 hectares of land). This is the harvesting time of rabi rice, sesame, gram, etc. Famers incurred losses due to non-availability of labor as well as lack of transportation facilities and market closure.

The District Purba Bardhaman is called the granary of West Bengal. Although Agriculture is the main activity of the selected districts, it is also the center of the development of industries like Iron and steel, coal mining, cement, oil mills, thermal power etc. Rice is the most important crop of the selected districts and in the alluvial plains to the east little else is grown. The rice grown with its numerous varieties can broadly be grouped under the three primary classes distinguished from one another by distinct characteristics: the aus or autumn, the aman (Fig. 1) or winter, and the boro or the summer rice. Paddy covers a maximum of the gross cropped area.

Before the lockdown, the majority of people in the study areas were self-sufficient and largely depended on agriculture, business and service for their livelihood. After the COVID-19 major portion of the workforce lost their income. The others continued their occupation remain more or less the same. Following the lockdown, many farmers and fishermen, daily wagers, petty traders were left jobless as most of the markets were closed and disrupted transportation (Muhammad et al., 2020).

Several reports showed that needy, vulnerable, poor, and far behind in receiving relief in cash or kind. NGOs and social workers were working in providing needy support. Therefore, the government should take precautionary measures at the grassroots level (Bhavani, 2020).

Farmers faced problems during the transport of their products and there was a lack of communication facilities to bring the crop to the local market, and most of them had lost jobs and income due to the lockdown. The rabi crop stood ready for harvest in many fields when the COVID-19 crisis brought everything to a halt; this is also the time for the harvest of plantation crops like pepper, coffee, banana. In the aftermath of the lockdown, the harvest of the rabi crops has been delayed due

Fig. 1 Farmers are planting aman paddy

to the non-availability of labor, machinery (harvesters, threshers, tractors), transport facilities and restrictions on movement; farmers of perishable commodities like fruits, vegetables, and flowers, in particular, have incurred losses. This lockdown has coincided with the time of harvesting of rabi crops and has hit farmers and allied both, directly and indirectly, due to lack of transport/ communication facilities and unavailability of the labor force (Fig. 2) and blockages to transport routes are particularly obstructive for fresh food supply chains and may also result in increased levels of food loss and waste (Siche, 2020). Fresh vegetables, fish, etc., which are highly perishable and therefore need to be sold, processed, or stored in a relatively limited time, are at particular risk. Farmers have adopted some strategies to cope with this lockdown. One such strategy is that farmer himself (Fig. 2) with the neighbor of the villagers who acted as a farmer sharing their labor themselves i.e., harvest their crop eventually, and they get benefitted with each other. Secondly, some farmers together contract a van/ car puller to drop their product to the nearby market. To date, most of the respondents agree that the government does not provide any compensation for the failure of crops. It is commonly accepted that farmers cannot work from home.

From the Table 1, it is amply evident that 50 percent of the respondents belonged to the middle age group, followed by old age (30.00%) and young (20.00%).

Fig. 2 A 75 yrs old himself carry rice because of lack of labour

Table 1 Distribution of the respondents according to age (n = 400)

Sl. no	Category	Number	Percentage
1	Young (up to 35 yrs)	80	20.00
2	Middle age (36 – 50yrs)	200	50.00
3	Old (above 50 yrs)	120	30.00

2 Occupation

Occupation is an important element indicating the socio-economic status of a family. It also indicates a major source of income for an individual for livelihood. The occupational background of an individual greatly influences his involvement in the various program implemented by the state developmental departments. Occupational distributions of the respondents were collected, analyzed and results have been depicted in the Table 2.

Table 2 Distribution of respondents according to their occupation (n = 400)

Sl. No	Occupation	Frequency	Percentage
1	Agriculture	100	25.00
2	Daily wages	100	25.00
3	Service	100	25.00
4	Business	100	25.00

Table 3 Distribution of respondents according to annual family income (n = 400)

Sl. No	Level of income (in Rs.)	Number	Percentage
1	Below 50,000	75	18.75
2	50,000–1,00,000	115	28.75
3	1,00,000–2,00,000	60	15.00
4	Above 2,00,000	150	37.50

3 Annual Family Income

The income of a family greatly influences the decision-making habit of an individual and family. The goal and aspirations of an individual are more or less ascertained based on his average annual income. It is very difficult to assess the average annual income as the respondents neither keep any records nor express correctly from memory. The researcher, therefore, made sufficient probing on the activities undertaken and income generated from each activity.

Table 3 depicts that most of the respondents' (37.50%) annual income comes above 2, 00,000 and 28.75 percent of the respondent generate annual revenue between 50,000 and 1, 00,000 followed by 18.72 percent belongs to below 50,000 income category.

4 Livelihood

Livelihood means how people secure their life. It comprises assets, capabilities, income and activities. A livelihood is sustainable when people can cope with and recover from calamities, pandemics and enhance their well-being and that of future generations without undermining the natural environment or resource base.

Table 4 depicts the impact of livelihood in the study areas. Before the lockdown, the majority of people in the study areas were self-sufficient and largely depended on agriculture, business and service for their livelihood. After the COVID-19 major portion of the workforce lost their income. The others continued their occupation remain more or less the same. Following the lockdown, many farmers and fishermen, daily wagers, petty traders were left jobless as most of the markets were closed and disrupted transportation. It is evident from the above information that major portions of the workforce engaged in daily wages laborers. A major portion of the

Table 4 Impact on livelihood

Sl. No	The main source of income	Before COVID 19 (%)	After COVID 19 (%)
5	Agriculture	25.00	20.00
6	Daily wages	25.00	10.00
7	Service	25.00	22.00
8	Business	25.00	10.00
9	Jobless	00.00	38.00

workforce migrates to nearby districts even to outside states for gainful employment. But because of lockdown, they are not able to go to the workplace. However, Mahatma Gandhi National Rural Employment Programme has created some scope for creating additional employment for job seekers at the Village level. People migrated to nearby towns or cities to search for jobs as the only alternative for livelihoods (Mondal et al., 2018). Most cultivators can cultivate only one crop a year. Many have become wage laborers and have lost secure jobs. The womenfolk, back at home, have to do all the work on the field and home.

COVID-2019 is affecting agriculture in two significant aspects: the supply and demand for food (WHO, 2020). These two aspects are directly related to food security, so food security is also at risk (FAO, 2020a).

5 Food Supply

The food supply chain is a network that connects an agricultural system (the farm) with the consumer's, consisting of a process that includes manufacturing, packaging, distribution, and storage (Chen et al., 2020).

Primarily, the announcements of social isolation forced people to go to the supply centers and generate a scarcity of some products. Regardless of this, the food supply has stabilized because it is one of the structures that should be maintained to ensure food safety. One of FAO's roles is to promote that food value chains are not interrupted and continue to operate (FAO, 2020b). Thus, despite the restrictions that governments have imposed on the mobility of labor in agricultural systems, although with some problems, the supply of basic necessities is normally assured. The situation is different when it comes to imported or exported goods; due to the closure of borders, international trade was interrupted, although after having defined security protocols to avoid the spread of the virus, trade stabilized. This may be temporary; it depends on what countries are doing to stop the spread of the virus.

Part of the food supply system is the social programs that some countries, mainly Latin America, have to feed millions of families and children with limited economic resources. This supply system is being served in different ways:

1. Distribution of basic food through public distribution system i.e. rations of basic necessities.
2. Financial arrangement is equivalent to the cost of food rations of basic necessity.

According to Zhang (2020), China experienced a greater impact on livestock due to a shortage of access to animal feed and labor. Interruptions to food transfers are minimal, so the food supply remains stable.

Although it depends on the country and the measures that each one has adopted, globally, the prices have remained stable. Therefore, no spikes in the prices of basic necessities are expected, although it is more likely to occur for high-value products, especially perishables, dairy and meat products. The FAO Food Price Index (FFPI) is indices that measure the worldwide price variation. According to the FAO (2020c), the FFPI of February 2020 had an average of 180.5 points, that is, 1.9 points (1.0%) less than in January, constituting the first month-on-month decrease after four months of consecutive increases.

Many governmental and non-governmental organizations worldwide are playing a crucial role in maintaining a continuous food supply chain to overcome these challenges. At present, Food and Agriculture Organization (FAO) is working with the main aim to maintain the food value chain and keep the food supply going (FAO, 2020a, 2020b, 2020c).

From Table 5, most of the farmers agreed that they faced problems during the transport of their product. There was a lack of communication facilities to bring the crop to the local market. Most of them had lost jobs and income due to lockdown. The rabi crop stood ready for harvest in many fields when the COVID-19 crisis brought everything to a halt; this is also the time for the harvest of plantation crops like pepper, coffee, banana. In the aftermath of the lockdown, the harvest of the rabi crops has been delayed due to the non-availability of labor, machinery (harvesters, threshers, tractors), transport facilities, and restrictions on movement; farmers of perishable commodities like fruits, vegetables, and flowers, in particular, have incurred losses. This lockdown has coincided with the time of harvesting of rabi crops and has hit farmers and allied both, directly and indirectly, due to lack of transport/communication facilities and unavailability of the labor force (Figs. 2 and 3). Blockages to transport routes are particularly obstructive for fresh food supply chains and may also result in increased food loss and waste levels. Fresh vegetables, fish, etc. which are highly perishable and therefore need to be sold, processed, or stored in a relatively limited time, are at particular risk. Farmers have adopted some strategies to cope with this lockdown. One such strategy is that farmer himself (Fig. 2) with the neighbor of the villagers who acted as a farmer sharing their labor themselves, i.e., harvesting their crop eventually, they benefit from each other. Secondly, some farmers together contract a van/ car puller to drop their product to the nearby market. To date, most of the respondents agree that the government does not provide any compensation for the failure of crops. It is commonly accepted that farmers cannot work from home.

Daily wagers include the van, car puller, migrant laborers, vegetable sellers, Mahatma Gandhi National Rural Employment Guarantee Programme workers, etc.

Table 5 The impact of lockdown on Farmers (n = 100)

Sl. No	Nature of assistance	Extent of opinion					Mean score	Rank
		Strongly agree	Agree	Undecided	Disagree	Strongly disagree		
1	The government provide sufficient measures to protect your family	22	26	25	16	11	3.32	V
2	Problems faced during transport/ communication facilities	45	30	25	00	00	4.20	I
3	Loss of jobs/income	45	43	00	12	05	4.06	II
4	The government provide compensation due to loss of jobs/income	00	17	00	48	35	1.99	IX
5	Able to purchase your basic food essentials at this time	12	19	00	41	28	2.66	VIII
6	Get a remunerative price with your products/ talent	00	00	00	63	37	1.63	X
7	Permanently shut down the job because of this crisis	11	13	45	11	20	2.84	VII
8	Continue to do your job after the lockdown	20	23	17	23	17	3.21	VI
9	Adopted some strategies to cope with the crisis	41	32	12	10	05	3.94	III

(continued)

Table 5 (continued)

Sl. No	Nature of assistance	Extent of opinion					Mean score	Rank
		Strongly agree	Agree	Undecided	Disagree	Strongly disagree		
10	Pandemic affected the ability to purchase inputs for your enterprise	15	55	00	12	18	3.37	IV

(Maximum obtainable score-5)

Fig. 3 Boro rice still in the field because of lack of labour

They were most affected during the lockdown and reported being unable to purchase inputs for their own job. They are the most vulnerable in the community. They do not have enough money for the future even though the government cannot provide compensation to the victims, in their opinion. After the lockdown, many are ready to do work for their livelihood. They have no option to generate their income as a whole to their livelihood. Workers/laborers are unable to go to their workplace due to lockdown in all sectors such as transport/ communication facilities. All the respondents are permanently shut down their job because of this crisis. Agriculture laborers are not able to go to work due to lack of transport. Labour work under the Mahatma Gandhi National Rural Employment Guarantee Scheme (MNREGS) has stopped. Migrant laborers who have managed to return will not be able to go back immediately. From Table 6, some may not wish to go back also, given the long journey many may have endured following the lockdown. There is a need for both relief and rehabilitation measures to help the affected pick the threads again, overcome the sustained loss, and rebuild their lives. Precautionary measures like maintaining social distancing and hand washing will have to be continued.

The lifeblood for economic growth and jobs, the service sector contracted in March as new business and export demand fell sharply as the Coronavirus pandemic wreaked havoc globally. Service holders comprise people engaging the public or private sector, banking sector, or those with a secure monthly income. Due to lockdown, thirteen

Table 6 The impact of lockdown on daily wagers (n = 100)

Sl. No	Nature of assistance	Extent of opinion					Mean score	Rank
		Strongly agree	Agree	Undecided	Disagree	Strongly disagree		
1	The government provide sufficient measures to protect your family	24	27	13	20	16	3.23	VII
2	Problems faced during transport/communication facilities	48	52	00	00	00	4.48	IV
3	Loss of jobs/income	51	49	00	00	00	4.51	III
4	The government provide compensation due to loss of jobs/income	12	17	00	41	30	2.42	IX
5	Able to purchase your basic food essentials at this time	13	21	00	53	13	2.68	VIII
6	Get a remunerative price with your products/talent	00	00	00	50	50	1.50	X
7	Permanently shut down the job because of this crisis	90	10	00	00	00	4.90	I
8	Continue to do your job after the lockdown	27	16	55	02	00	3.68	V
9	Adopted some strategies to cope with the crisis	27	32	12	15	24	3.53	VI
10	Pandemic affected the ability to purchase inputs for your enterprise	54	46	00	00	00	4.54	II

(Maximum obtainable score-5)

people have lost their income. They strongly agreed that transport/communication facilities are a burden to join the office at this time. Some are doing their job work from home and had time to do things that a busy lifestyle did not permit. 'Loss in income and jobs and cautiousness among consumers will delay recovery in consumer demand even after the pandemic' (PTI). The Table 7 has seen a decrease in the income of high-paid people; most of them have permanent jobs and are also able to work from home during the lockdown. The availability of food and other items is less than usual, but very few respondents indicated that key items were unavailable. The staple food, fresh food and hygiene items were generally always or sometimes available in stores, with some of the respondents stating that they were unavailable. Fresh food appears marginally less available than staples (Principato et al., 2020). Noticeably fewer respondents indicated that medicines were always available, with one-fifth not knowing either way.

Many businesses struggled, and some may even die. But as with economic adversities of the past, new industries will emerge, bringing with it renewed hope of recovery.

Table 7 The impact of lockdown on Service sectors (n = 100)

Sl. No	Nature of assistance	Extent of opinion					Mean score	Rank
		Strongly agree	Agree	Undecided	Disagree	Strongly disagree		
1	The government provide sufficient measures to protect your family	22	18	00	50	10	2.92	V
2	Problems faced during transport/communication facilities	40	32	00	10	18	3.66	I
3	Loss of jobs/income	05	08	00	42	45	1.86	VII
4	The government provide compensation due to loss of jobs/income	00	00	15	48	37	1.78	VIII
5	Able to purchase your basic food essentials at this time	00	13	00	37	50	1.76	IX
6	Get a remunerative price with your products/talent	10	09	12	33	39	2.27	VI
7	Permanently shut down job because of this crisis	00	7	5	33	55	1.64	X
8	Continue to do your job after the lockdown	27	36	00	14	23	3.30	II
9	Adopted some strategies to cope with the crisis	31	18	00	27	24	3.05	IV
10	Pandemic affected the ability to purchase inputs for your enterprise	30	22	00	24	24	3.10	III

(Maximum obtainable score-5)

Eventually, things will go back to normal, just that we'll have to change the definition of normal. Business and industry will get hit adversely due to disruption in labor supply, raw material unavailability, working capital constraints, and restricted demand due to limited movement of people and purchasing ability. Working at home is not possible for small business holders and they are not able to go to their business place to run their job. They sell their products which were stored, but consumers were not remarkable. So, as a result, they lose their job. Some reported that they are not continuing their small business after the lockdown. Table 8 showed that some have left their job and doing some daily wage operations for survival. The micro, small and medium enterprises as a whole form a major chunk of manufacturing in India and play an important role in providing large-scale employment and contributing to the country's exports. Although the pandemic affected all businesses, the MSME (Micro Small and Medium Enterprise) sector would be badly hit by reduced cash flows caused by the nationwide lockdown. Their supply chain would be disrupted, and they would be affected by the exodus of migrant workers, restrictions in the

Table 8 The impact of lockdown on Businesses (n = 100)

Sl. No	Nature of assistance	Extent of opinion					Mean score	Rank
		Strongly agree	Agree	Undecided	Disagree	Strongly disagree		
1	Govt provide sufficient measures to protect your family	13	17	00	37	33	2.40	VIII
2	Problems faced during transport/communication facilities	38	50	00	12	00	4.14	I
3	Loss of jobs/income	15	19	00	31	35	2.48	VI
4	The government provide compensation due to loss of jobs/income	00	00	00	45	55	1.45	X
5	Able to purchase your basic food essentials at this time	13	17	00	38	32	2.41	VII
6	Get a remunerative price with your products/talent	04	10	00	46	40	2.12	IX
7	Permanently shut down job because of this crisis	13	12	52	09	04	3.31	IV
8	Continue to do your job after the lockdown	21	22	00	38	19	2.88	V
9	Adopted some strategies to cope with the crisis	35	38	00	12	15	3.66	III
10	Pandemic affected the ability to purchase inputs for your enterprise	43	36	00	13	08	3.93	II

(Maximum obtainable score-5)

availability of raw materials, the disruption to exports and imports, and also by the widespread travel bans, closure of malls, hotels, theatres, and educational institutions, etc (Padhee and Carberry, 2020). This, in turn, would massively hamper the MSME businesses. As a consequence, hundreds of thousands of people who work for these small businesses may end up with job and salary losses.

6 Coefficient of Concordance

Consider Table 9, which contains illustrative data. In Friedman's test, the null hypothesis is that there is no real difference among the n objects (assistance, rows of Table 9) because they pertain to the same statistical population. Under the null hypothesis, they should have received random ranks among the various variables so that their sums of ranks should be approximately equal. Kendall's test focuses on their variables. If the null hypothesis of Friedman's test is true, this would mean that the variables have produced rankings of the objects that are independent of one another. This is the null hypothesis of Kendall's test. (Salkind, 2010).

Four organizations rank order the same assistance with the results shown in Table 9. The average rank is used in cases of ties. Calculate Kendall's W for this data and test whether there is no agreement among the stakeholders.

Here, m = 4, n = 10

Putting this value in the above equation, get

$$\text{Kendall W} = 0.6727$$

Kendall's W represents an agreement, where 0 means no agreements at all, and 1 represents perfect agreement. Although the coefficient of 0.67 indicates some level of agreement.

$$\text{ChiSq} = 24.2182\,\text{df} = 9\,\text{p} = 0.004$$

Testing the Significance of W (Salkind, 2010).

Friedman's chi-square statistic is obtained from W by the formula.

$$\chi^2 = m(n-1)W.$$

This quantity is asymptotically distributed like chi-square with v = (n-1) degrees of freedom; it can be used to test W for significance. According to Kendall and Babington Smith, this approach is satisfactory only for moderately large values of m and n. Here, df = (10–1) = 9

$$\textbf{\textit{Chi square}}\,(\chi^2) = 24.2182$$

Table 9 Rank position

Sl. No	Assistance	Stakeholder				Sum
		Farmers	Daily wagers	Service holders	Business holders	
1	Govt provide sufficient measures to protect your family	5	7	5	8	25
2	Problems faced during transport/ communication facilities	1	4	1	1	07
3	Loss of jobs/income due to Covid-19	2	3	7	6	18
4	The government provide compensation due to loss of jobs/income	9	9	8	10	36
5	Able to purchase your basic food essentials at this time	8	8	9	7	32
6	Get remunerative price with your products/talent	10	10	6	9	35
7	Permanently shut down the job because of this crisis	7	1	10	4	22
8	Continue to do your job after the lockdown	6	5	2	5	18
9	Adopted some strategies to cope with the crisis	3	6	4	3	16
10	Pandemic affected the ability to purchase inputs for your enterprise	4	2	3	2	11

(continued)

Table 9 (continued)

Sl. No	Assistance	Stakeholder				Sum
		Farmers	Daily wagers	Service holders	Business holders	
SUM		55	55	55	55	

We see that $W = 0.6727$, which indicates some level of agreement among the respondents. We also see that and that the p-value $= 0.004 < 0.05 = \alpha$, thereby allowing us to reject the null hypothesis that there is no agreement among the respondents on the impact of lockdown on their livelihood.

7 Strategy to Improve Food Productivity

E-governance is the application of information and communication technologies to transform the efficiency, effectiveness, transparency and accountability of informational and transactional exchanges within government, between government and government agencies of National, State, Municipal and Local levels, citizen and businesses, and to empower citizens through access and use of information (Shah, 2007).

Some agricultural development services that can be provided in the developing world using ICT, are:

- Online services for information, education and training, monitoring and consultation, diagnosis and monitoring, and transaction and processing;
- E-commerce for direct linkages between local producers, traders, retailers and suppliers;
- Question-and-answer services where experts respond to queries on specialized subjects ICT services to block and district level developmental officials for greater efficiency in delivering services for overall agricultural development;
- Up-to-date information, supplied to farmers as early as possible, about subjects such as packages of practices, market information, weather forecasting, input supplies, credit availability, etc.;
- Creation of databases with details of the resources of local villages and villagers. Site-specific information systems, expert systems, etc.;
- Provision of early warning systems about disease/pest problems, information about rural development programs and crop insurances, post-harvest technology, etc.;
- Facilitation of land records and online registration services;
- Improved marketing of milk and milk products;
- Services providing information to farmers regarding farm business and management;
- Increased efficiency and productivity of co-operative societies through the computer communication network and the latest database technology;

- Tele-education for farmers;
- Websites established by agricultural research institutes, making the latest information available to extension (knowledge) workers and obtaining their feedback; and
- Mobile-based apps on the package of practices, market information, market linkages, weather, disaster management, crop protection, crop insurance, credit facilities, details of public and private sector officers dealing with agriculture and related sectors.

8 Use of Information Technology

The Information Technology revolution is unfolding and has very high visibility. However, its benefits have remained confined primarily to the urban areas, and rural communities have not been able to gain to the same extent from IT. As a means of agricultural technology transfer to farmers, information technology has had a limited impact. Even the vast potential of the broadcasting network has been tapped only minimally for an extension.

Increasing Use of Information Technologies. Harnessing Information Technology for agricultural extension will receive high priority in the new policy agenda. Extensive use of modern information technology will be promoted for communication between researchers, extension workers and their farmer clients to transfer technologies and information more cost-effectively. Information technology should be made available to those with specific inquiries to guide them in adopting the more knowledge-intensive forms of agriculture, which will expand in the future (Chang and Xu, 2021).

IT Application in Agriculture Marketing. Agriculture produces marketing requires connectivity between the market and exporters/growers/traders, industry consumers, through a wide area network of national and international linkages to provide day-to-day information with regard to commodity arrivals and prevailing rates, etc. to provide links for online International market information; to provide export-related documentation, to inform about the latest research in agricultural marketing, packaging, storage related information and to provide connectivity with lead international and national market organizations.

Wider Use of Electronic Mass Media for Agricultural Extension. Radio and TV have vastly increased their reach, as also reception facilities. "Local" radio and new FM transmitters open up possibilities of area-specific broadcasts. In communicating with an audience with low literacy skills, an audio-visual medium like TV has advantages. Today Doordarshan covers the entire population. Much wider and creative use of the mass media—All India Radio, private FM, Doordarshan, the private cable network will be promoted for more rapid and effective dissemination of general information and advice to farming communities.

Farmer Participation in IT Programmes. In developing any system of IT for agriculture technology transfer, the farmer is to be kept center-stage. She/he is not to be treated as a passive recipient but rather as a player, a generator and user of knowledge. The upgrade of his/her skills and knowledge is, therefore, a crucial part of the process. The farmer will be an effective participant in the process.

Support to States for Information Technology. Increased use of information technology at State/ district and block levels would be promoted. This would include electronic access, through NICNET, to technical and administrative information. Central Government will support States in the use of electronic linkages and computerization so that marketing, research, extension and farming communities are linked to each other and into local, national and global networks.

Private Information Shops/Kiosks. The ultimate aim is to promote private Information shops/ kiosks franchised out to the private sector, especially unemployed rural educated youth, in the manner of PCOs/STD shops. The private sector will be encouraged to establish information shops at the Block/Mandal/Village level. A major software development program will need to be mounted so that Information Shops can have access to suitable material. Electronic connectivity and access to e-mail would put the franchisees in contact with district KVKs, Line Departments, markets, and other sources of information. Such information could be dispensed to farmers, farmers groups on payment.

Capacity Building for the use of IT: Application of IT is constrained by lack of or inadequacy of complementary inputs (equipment, power, etc), appropriate organizational and institutional structures, information management and skills development. A major training program for developing capacity for IT usage will be promoted. Training Institutes will run suitable courses for the purpose.

9 Agricultural Knowledge and Information System (AKIS)

Agricultural Knowledge and Information Systems (AKIS) comprise the institutions and organizations that generate and disseminate knowledge and information to support agriculture production, marketing, and post-harvest handling of agricultural products and management of natural resources. Most AKIS projects support agricultural research, extension, or education activities, increasingly viewed as inter-related system components.

An AKIS is a system that links rural people and institutions to promote mutual learning and generate, share and utilize agriculture-related technology, knowledge and information. The system integrates farmers, agricultural educators, researchers and extensionists to harness knowledge and information from various sources for better farming and improved livelihoods. This integration is suggested by the "knowledge triangle".

AKIS systems that are financially, socially and technically sustainable; relevant and effective processes of knowledge and technology generation, sharing and uptake; AKIS systems that are demand-driven through the empowerment of farmers such that programs and activities are responsive to their needs; the interface between and

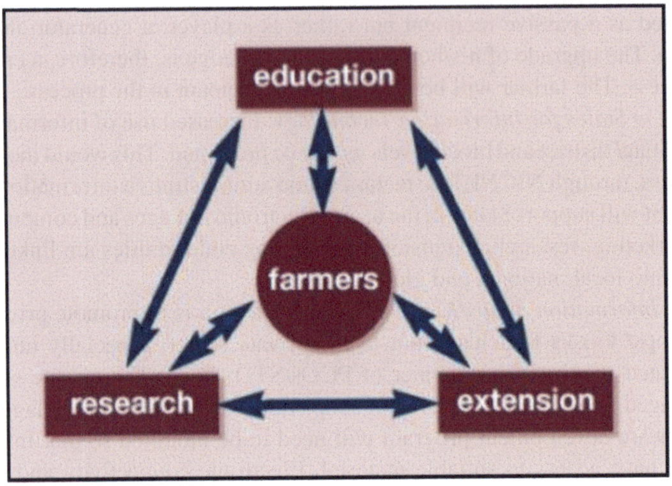

Fig. 4 Model of AKIS. *Source* Sharma and Hasan (2012)

integration among the various education, research, extension and farming activities; and accountability to assure that stakeholders assume their respective responsibilities.

The AKIS concept and practice hold significant promise for the advancement of agricultural and rural development and, more generally, national economies. In order to realize the value and importance of the AKIS concept, agricultural institutions need actively to promote linkages, technology transfer, knowledge sharing and the exchange of relevant information and such an impetus to the development of pluralistic innovation systems must be supported by adequate financial commitment. Fundamental to the development of an AKIS is recognition of the role of a plethora of private sector actors (seed and input supply companies, produce buyers, chemical companies, radio and television, etc.) playing different roles within the system.

Rural people, especially farmers, are at the heart of the knowledge triangle. Education, research and extension are services—public or private—designed to respond to their needs for knowledge with which to improve their productivity, incomes and welfare and manage the natural resources on which they depend in a sustainable way. Shared responsiveness to rural people and an orientation towards their goals ensures synergies in the activities of agricultural educators, researchers and extensionists. Farmers and other rural people are partners within the knowledge system, not simply recipients.

An Emerging Approach for Sustainable Development (Saha, 2011) reported that there is a conceptual progression from looking at various institutions and practices such as farming system development, extension and research in isolation to considering the linkages between the pairs of these elements as an Agricultural Knowledge Information System. AKIS could involve providing farmers with a basket of opportunities and helping them to choose the right opportunity for their situation.

An idealized AKIS model: This model (Fig. 4) suggests the numerous elements in AKIS. In fact, it could be made messier in that the policy, physical and human resources, communications and institutional elements should be connected to each of the four main sets of institutions—research, extension, education and support systems—which themselves should include both public and private sector entities.

10 New Opportunities Exist for Raising AKIS Effectiveness

Advances in the Agricultural sciences are crucial but other advances are also needed.

(1) Communication and Information Technologies are advancing rapidly
(2) New concepts are emerging for participation in learning and problem solving

The government also needs to create the conditions necessary for developing AKIS. Investment in market development and support to input providers, especially credit and supply institutions, is needed to stimulate the agricultural community, and attention to the rural physical infrastructure is needed to make the environment attractive and safe. Agricultural producers, especially women and poor farmers, require education and training to bring them into the modern world of labor-saving technologies and more productive practices. Joint planning between producers and institutional operators can provide the platform for advancing a demand-driven system of technological innovation for agricultural development. System leaders and managers need a better understanding of the dynamic nature of both national and international technology systems and should be able to identify those areas where the public system has a comparative advantage over private-sector R&D firms. The development of AKIS is attractive to the private sector. Major roads that link towns will almost certainly have to be built or funded by the government. The benefits from roads will be widely spread and accrue as much to the public as to the private sector regarding access to clientele or potential clientele. In the short and—especially—the long terms, private sector issues of distribution and dynamic efficiency promise to be enhanced as a result of government commitment to AKIS.

11 Strategy to Strengthen the Agricultural Sector Post-COVID-19

1. Due to the decline in agriculture and poultry production and productivity, the government should enhance income and livelihood opportunities through various schemes.
2. Sustainability is the option to strengthen the agricultural sectors. Farmers themselves can produce nutritious food. As a result, save biodiversity, develop new strategies, and develop innovations to tackle social, economic, and environmental challenges (FAO and IFAD, 2019).

3. Agriculture in combination with poultry, dairy and fisheries should heed the success in increasing productivity and profits. Unlike cereals, pulses and vegetables, milk prices were not affected by the lockdown (Cariappa et al., 2020a). These together will have the potential to overcome the challenges of production and marketing risks of the farmers.
4. Universalization of Mahatma Gandhi National Rural Employment Guarantee Scheme for migrant workers who have returned their home.
5. Set up infrastructure in rural areas so that farmers can store their products to storage during this lockdown.
6. MSMEs may be encouraged to produce PPE kits and hand sanitizers, etc.
7. Agricultural loan may be waived for at least one year may be thought of.
8. Providing market information through various social media.

12 Conclusion

The national lockdown from March 25, 2020, has severely affected lives and livelihoods across rural India. Livelihood in many ways disrupted due to lockdown. Vulnerable sections of our community are not able to sustain their livelihood. Foods, necessary commodities, medicines appeared to be less than the demand—respondents who have the financial ability stored a large number of food items for the uncertain future. Service holders are relatively in a secure place as they are salaried, but the rest of the people, such as pity traders, migrant workers, daily wagers, are at risk for their livelihood. Transportation was totally stopped from March 25, 2020. Widespread discontinuation of livelihoods is translated into the loss of income, particularly for vulnerable people such as casual labor or petty trade, business owners, and others. Some reported that they had to reduce their consumption or skip one/two meals a day to avoid food shortage. It is essential to mitigate the disruptions to livelihoods and cushion the financial blow. It is concluded from this study that respondents agreed among themselves that their livelihood had been shaken owing to lockdown and they are at risk to generate their livelihood. Therefore, the government should have a focus on people's life and livelihood. To date, the number of Corona patients has increased at an increasing rate. This study suggested that the availability of food, hygiene items, and medicines will appear to be normal. Lockdown should be properly implemented in such a way so that livelihood can't be hampered—building herd immunity among the people in the major and foremost important step to overcome the pandemic. The government should have to create alternative livelihood options or provide financial and food security to sustain their life and livelihood.

References

Alankar, B., Kaur, H., Ahsaan, S., Sharma, G., & Chang, V. (2021). Towards reviewing an immediate impact of COVID-19 on the integrative world economy: An evolving perspective. *Journal of Global Information Management (JGIM), 30*(4), 1–19.

Bhavani, R. V. (2020). Impact of COVID-19 on rural lives and livelihoods in India. *India Matters*. Retrieved April 20, from 2020.

Chang, V., & Xu, Q. (2021). Analysis of influencing factors of grain yield based on multiple linear regression. *International Journal of Business and Systems Research, 15*(3), 337–355.

Chen, S., Brahma, S., Mackay, J., Cao, C., & Aliakbarian, B. (2020). The role of smart packaging system in food supply chain. *Journal of Food Science, 85*(3), 517–525.

Dev, M. & Sengupta, R. (2020a). Covid-19: Impact on the Indian economy, Indira Gandhi Institute of development research, Mumbai.

Dev, M. & Sengupta, R. (2020b). Addressing COVID-19 impacts on agriculture, food security, and livelihoods in India. *IFPRI Blog*.

FAO. (2020c). FAO Food Price Index. Retrieved from http://www.fao.org/worldfoodsituation/foodpricesindex/en/.

FAO (2020b). FAO Director-General urges G20 to ensure that food value chains are not disrupted during COVID-19 pandemic.

FAO. (2020a). Q&A: COVID-19 pandemic - impact on food and agriculture. Retrieved from http://www.fao.org/2019-ncov/q-and-a/en/.

Mondal, D., Chowdhury, S., & Basu, D. (2018). Role of panchayat (Local self-government) in managing disaster in terms of reconstruction, crop protection, livestock management and health and sanitation measures. *Natural Hazards, 94*(1), 371–383.

Muhammad, S., Long, X., & Salman, M. (2020). COVID-19 pandemic and environmental pollution: a blessing in disguise? *Science of the Total Environment* 138820.

Padhee, A. K., & Carberry, P. (2020). Containing COVID-19 impacts on Indian agriculture: International crops research institute for the semi-arid tropics (ICRISAT). *Agriculture Times*.

Principato, L., Secondi, L., Cicatiello, C., et al. (2020). Caring more about food: The unexpected positive effect of the Covid-19 lockdown on household food management and waste. *Socio-Economic Planning Sciences*. https://doi.org/10.1016/j.seps.2020.100953

Saha, A. (2011). Natural resource management – An emerging domain in sustainable development. In D. Das Gupta (Ed.), *Environmental Concerns*, pp. 127–132, ISBN No. (10): 81-7754-428-4. Agrobios, Jodhpur, India.

Salkind, N. J. (Ed.). (2010). *Encyclopedia of research design* (Vol. 1, pp. 164–167). Sage.

Sharma, A., & Hasan, S. (2012). Information communication technologies usage by undergraduate students in Pantnagar University. *Interaction, 30*(1), 97–101.

Siche, R. (2020). What is the impact of COVID-19 disease on agriculture? *Scientia Agropecuaria, 11*(1), 3–6.

WHO (2020). World Health Organization. COVID-19 Situation update for the WHO African Region. External situation report. Retrieved from https://apps.who.int/iris/bitstream/handle/10665/331763/SITREP_COVID-19_WHOAFRO_20200415-eng.pdf.

Zhang, X. (2020.) Chinese livestock farms struggle under COVID-19 restrictions. Research post of international food policy research institute.

COVID-19 Patients Management and Triaging Using Machine Learning Techniques

Harleen Kaur, Iftikhar Alam, Ritu Chauhan, and Bhavya Alankar

Abstract COVID-19 is a disease that is caused by a new virus, coronavirus, which first appeared in China and a few months; it spread all over the globe, infecting many people. This disease shows very common symptoms like fever, cough, and tiredness, which makes it more difficult to know if the person is infected or not. There have been a lot of struggles in finding a way to detect the virus in a human body and manage the infected at the same time. There is an immense increase in the number of infected cases, so it becomes difficult to manage patients with proper resources and medical facilities, leading to an increase in casualties. To overcome the difficulty, this study proposes fast and efficient methods for the detection of the virus and proper treatment. COVID-19 patient management and triaging means accurately identifying patients or detecting COVID-19 and categorizing the patients or sorting them accordingly for their proper management. This study aims to help the government and health care system take relevant steps to detect and manage COVID-19 patients. Also, with the details and symptoms of the infected person, we can categorize the person as a mild, critical, or severe case. The proposed methods in the chapter have shown promised results while testing on COVID CT Scan Images and patients' symptoms dataset.

Keywords COVID-19 · Machine learning · Convolutional Neural Network (CNN) · Image classification · Triaging

---

H. Kaur (✉) · I. Alam · B. Alankar
Department of Computer Science and Engineering, School of Engineering Sciences and Technology, Jamia Hamdard, New Delhi, India
e-mail: harleen@jamiahamdard.ac.in; harleen.unu@gmail.com

I. Alam
e-mail: iftikharalam97@gmail.com

B. Alankar
e-mail: balankar@jamiahamdard.ac.in; bhavya.alankar@gmail.com

R. Chauhan
Centre for Computational Biology and Bioinformatics, Amity University, Noida, U.P, India
e-mail: rituchauha@gmail.com

V. Chang et al. (eds.), *Artificial Intelligence and Machine Learning Methods in COVID-19 and Related Health Diseases*, Studies in Computational Intelligence 1023,
https://doi.org/10.1007/978-3-031-04597-4_10

1 Introduction

The disease has reached almost every part of the world and the number of infected cases is increasing day by day. It mostly affects the people who belong to the old age group or those who have less immunity. Such a situation has made an urgent call for a solution. The COVID-19 pandemic has affected individuals and organizations in several ways around the world. The machine learning approach has been widely used in the past, and the effective use of the machine learning approach in the pandemic is required across different sectors (Zhang et al., 2020). The machine learning approach in healthcare can be used in identifying, detecting, or predicting the positive and negative cases, thereby managing and controlling its spread, accelerating the development of new vaccines and drugs, and many others. Frequent steps have been taken in every country to treat the infected people, provide medical facility and quarantine the patients to stop the further spread of the disease (Wang, 2020). As the number of infected increases, it is becoming difficult for the medical department to manage the patients properly. And researchers and scientists are working on finding a way out of this Abbas et al. (2020), Alankar et al. (2021), Chang et al. (2021a). Many detection techniques came into light and computerized tomography (CT) scans are considered one of the best detection methods, as it is one of the advanced techniques, Narin et al. (2020), Abdel-Basset et al. (2021a, b), Chang et al. (2021b). We will be using the computerized tomography (CT) scan images for this study. Figure 1 shows a CT scan of an infected and non-infected person.

This study uses the deep learning technique because many deep learning models have been proposed to deal with image classification problems (Greenspan et al., 2016; Kaur et al., 2021). It is a convolutional neural network that is used to work with image datasets. A neural network is built using different layers, including convolutional layers and fully connected layers, Ge'ron (2019). Convolution helps in condensing the image and bringing out the more efficient features of the image. And this is done by passing the image through a filter, which helps change the image size

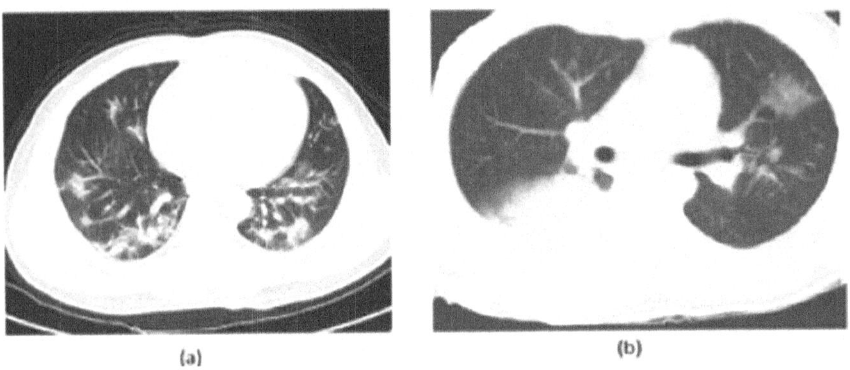

Fig. 1 **a** Image of COVID-19 case, **b** Image of normal case

and emphasize some of its features (Apostolopoulos and Mpesiana, 2020). Image Generator is an important technique to generate batches of tensor image data with real-time data augmentation. In the end, we will be building the model, configuring the model, and then training the model on the training data and testing the data on the validation data. Then, we will be able to predict whether the CT scan image is of an infected person or not.

1.1 Related Works

In the last few months, researchers have used the deep learning technique to solve the problem. Even they have used the concept of transfer learning to make the model more accurate using the Convolutional neural network (CNN) models based on VGG16, ResNet50, and InceptionV3 (Khan & Aslam, 2020). One has made use of the Decompose, Transfer, and Compose (DeTraC) model, which consists of three phases- starting with training the CNN model and extracting features and then using sophisticated gradient descent optimization method on the model and finally refining the model, Abbas et al. (2020). Researchers have also used a tailored architecture model for detection from Chest X-Ray (CXR) images. They have also used some complex models as well, like the one that uses three components—a backbone network to extract features and a classification and anomaly head that generates some score to calculate the result, Wang, (2020). The authors, Gulsen (2020) have presented work on the Simple classification of COVID–19 patients. This simple classification can aid an appropriate initial categorization of COVID-19 patients that can be improved further based on various findings that emerge over time. They use a clustering model on clinical cases to get a reliable result. Thakur and Kumar (2021) have presented work on the Classification of COVID-19 patients from chest CT images using multi-objective differential evolution-based CNN. The study uses CNN to classify the COVID-19-infected patients as infected (+ve) or not (-ve) Lu X et al. (2020) have shown classification of COVID-19 in intensive care patients: towards rational and effective clinical triage. The model was effective in predicting the COVID-19 epidemic. In this study, they bridge the gap between disease classification and clinical outcome in intensive care patients, Han et al. (2020) and Foster et al. (2014). Their classification helps in the individual evaluation and provides effective triage for treatment and management (Sorjamaa et al., 2007; Kaur and Wasan, 2006).

2 Methodology

The study is divided into two phases: Patient Recognition and Patient Sorting. Patient Recognition includes detection of COVID-19 virus using the CT scan image dataset. We are building an image classification model using deep learning techniques for the detection of patients. The input of the model will be the CT scan image of the person and the output will be telling us whether the person is infected with the virus or not. Patient Sorting is necessary as we are limited with the resources and we must work in a manner to be more efficient. So, it will be very useful if we can sort the patients, in accordance of their emergency level, by using the patient's details such as their symptoms like breathing issues, drowsiness, chest pain, body temperature and age, and pre-existing diseases like diabetes of the patient and traveling history, etc. (Chauhan et al., 2021). It is a multi-class classification problem, in which we will be dividing the patients into mild, severe, and critical categories.

A. *Dataset Description*

The proposed model utilizes the COVID-CT Scan image dataset, which is taken from a public accessible GitHub repository (Charles, 2013). The COVID-CT dataset has 350 CT images containing clinical findings of COVID-19 and 350 Non-COVID CT images from 216 patients. We have taken a total of 600 images to consist of 300 COVID-19 and 300 Non-COVID-19 CT scans. For Patient sorting, we have taken a dataset of infected persons and it is taken from the Kaggle database from the Kaggle website (https://www.kaggle.com/allen-institute-for-ai/CORD-19-research-challenge). It includes all the relevant details of their symptoms and has 127 observations with 20 feature columns. The data is structured with no null or missing values.

B. *COVID-19 Detection Model*

(1) *Importing the Data:* The COVID-CT image dataset has two sub-directories: training and validation, which contains CT scan images of infected and non-infected people, as shown in Fig. 2. The training and validation set will be fed to the trained generator and the validation generator, respectively. The names of the sub-directory will act as labels for the images that are contained within them.

The first step of importing the data is defining the path to the training and validation directory, followed by creating the sub-directories for COVID-19

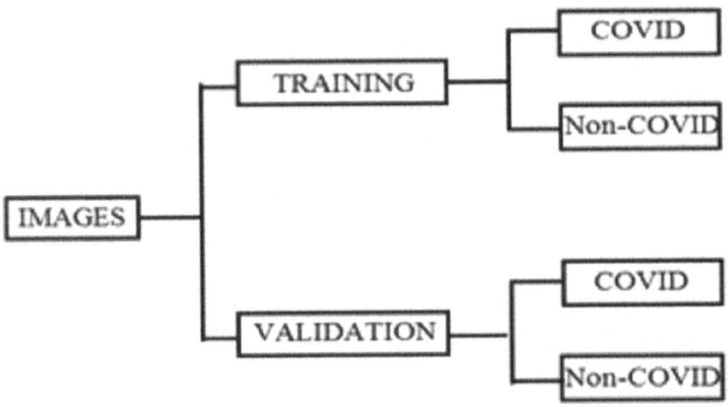

Fig. 2 Structure of image directory

and normal cases for both the training and validation directory. The proposed model uses 600 images for the training set and 100 for the validation set.

We can also see the input images by using the *matplotlib* library, by displaying a batch of COVID-19 and normal case images, fitting it to a size of (4 × 4). The images of COVID-19 and normal cases are shown in Figs. 3 and 4, respectively.

(2) *Building a Convolutional Neural Network (CNN) Model:* When dealing with an images dataset, we use a set of layers to build the model and they are convolutional layer, pooling layer, and fully connected layer. Starting with the Conv2D layers, this layer helps in condensing an input image by passing it to a filter and then using an activation function on the output vector (Lee & Ko (2009), (Yang et al. (2009). The arguments of the Conv2D layer are as follows:

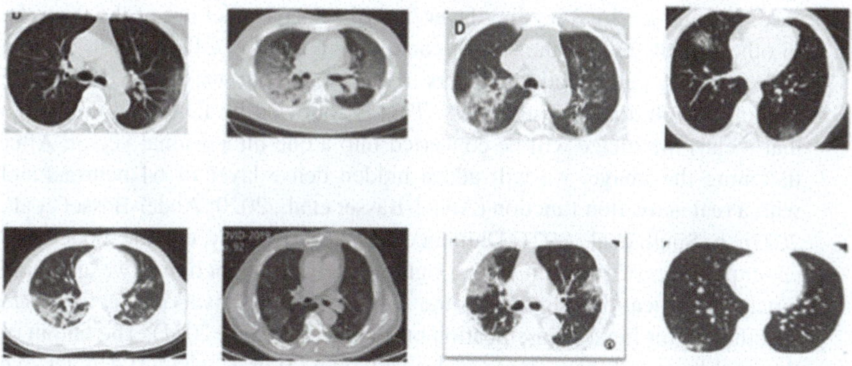

Fig. 3 Sample images of COVID-19

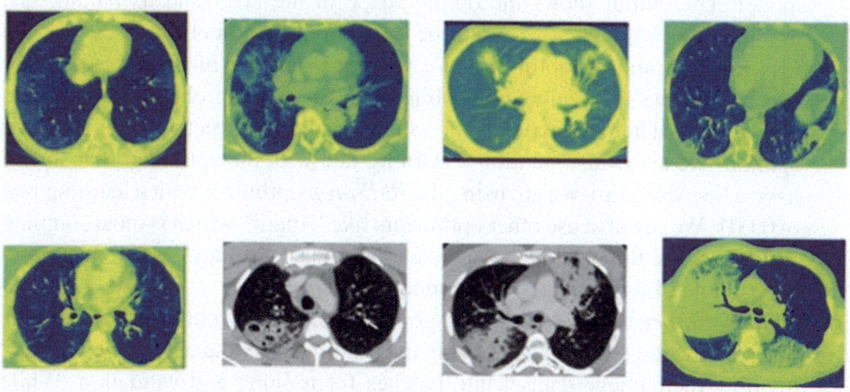

Fig. 4 Sample images of non-COVID-19

i. Filters: it is the number of outputs filters in the convolutions.
ii. Kernel size: denotes the height and width of the 2D convolution window. In our case, it is (2×2).
iii. Activation function: we will be using the *'relu'* activation function. Relu uses the following formula -If $x > 0$ return x, else return 0.
iv. Input shape: it mentions the size of the input image, in our case, the input shape = (224, 224, 3).

   Here 3 denotes the red, green, and blue channel, if the image is in grayscale, then it will be equal to 1.

A Convolutional layer is followed by a pooling layer *(tf.Keras.layers.MaxPool2D)*. This layer helps reduce the size of the image and keeps only the most effective elements of the image vector. We also must specify the size of the pooling layer we are going to use, which is 3×3 in our case. If using a 3 by 3 pooling layer, it will select 1 pixel out of the 9 pixels. In other words, reduce the size of each feature map by a factor of 3. We can add as many as convolutional layers in our model, but we will be using three sets of convolution-pooling layers. This is followed by flattening the image that means the image will be converted into a one-dimensional vector. After flattening the image, we will add a hidden dense layer of 64 neurons and with a real activation function (Abdel-Basset et al., 2020; Abdel-Basset et al., 2021a, b; Singh et al., 2021; Dhiman et al., 2021). Finally, end the model with an output layer of 1 neuron, with a sigmoid activation function as we are doing binary classification. Sigmoid is the most common activation function used specifically for binary classification problems (Tai et al., 2021). The output of the final layer will be a single scalar value, i.e., 0 or 1, where 0 denotes one class and one the other class.

The command *model.summary()*, provides the output of each layer of the model. The output shows the output shape of the layer and its parameters, as shown in Fig. 5. Next, we will be compiling the model by specifying the loss function and an optimizer. We will be using the *binary cross-entropy* loss because we are dealing with a binary classification problem. Optimizer is something that helps in updating the weights of the neural network in backpropagation. We also must mention a learning rate with the optimizer that should have a low value. So, we are using the *RMSprop* optimizer with a learning rate of 0.001. We can also use other optimizers like 'Adam', which is most common and will work fine in this case. We can add the 'accuracy' as the metrics to keep a note of the accuracy of the model while training.

(3) *Data Pre-processing:* Next, we are going to pre-process our data that includes normalizing the image data, defining the training and validation generator, and dividing their image dataset into batches for training and validation. While defining the training and validation generator, we must mention the training and validation dataset source, respectively. But first, we must normalize our input images, as we cannot feed raw pixels to the model. The image pixels values range from 0 to 255, so we can change their value to between 0 and 1,

```
Model: "sequential"
_________________________________________________________________
Layer (type)                 Output Shape              Param #
=================================================================
conv2d (Conv2D)              (None, 223, 223, 16)      208
_________________________________________________________________
max_pooling2d (MaxPooling2D) (None, 74, 74, 16)        0
_________________________________________________________________
conv2d_1 (Conv2D)            (None, 73, 73, 32)        2080
_________________________________________________________________
max_pooling2d_1 (MaxPooling2 (None, 24, 24, 32)        0
_________________________________________________________________
conv2d_2 (Conv2D)            (None, 23, 23, 64)        8256
_________________________________________________________________
max_pooling2d_2 (MaxPooling2 (None, 7, 7, 64)          0
_________________________________________________________________
flatten (Flatten)            (None, 3136)              0
_________________________________________________________________
dense (Dense)                (None, 64)                200768
_________________________________________________________________
dense_1 (Dense)              (None, 1)                 65
=================================================================
Total params: 211,377
Trainable params: 211,377
Non-trainable params: 0
_________________________________________________________________
```

Fig. 5 Summary of the proposed model

by using the 'rescale' function of the image data generator. Next, we will be defining the train and validation generator using *flow from directory()* function and mentioning the source path, with an image size of 224 × 224, train batch size of 60, and validation batch size of 10, as shown in Fig. 6. The batch size depends on the number of input images and the number of epochs we are using to train our model and finally mentioning the class mode as 'binary', as we are doing a binary classification problem.

(4) *Training:* Now, we can train our model. It can be trained for any number of epochs until its loss is decreasing. We are training our model using the *model.fit* method for 20 epochs, as shown in Fig. 7. We must mention the training and validation data in the function.

C. COVID-19 Patient Sorting Model

(1) *Exploratory Data Analysis:* The dataset has all the continuous data values, making it easier for us to build a machine learning model. The dataset has 20 columns with 19 independent variables that include—age, gender, body temperature, Dry Cough, sour throat, weakness, breathing problem, drowsiness, pain in the chest, travel history to infected countries, diabetes, heart

```
from tensorflow.keras.preprocessing.image import ImageDataGenerator

# All images will be rescaled by 1./255
train_datagen = ImageDataGenerator(rescale=1/255)
validation_datagen = ImageDataGenerator(rescale=1/255)

# Flow training images in batches of 128 using train_datagen generator
train_generator = train_datagen.flow_from_directory(
        '/content/drive/My Drive/Colab Notebooks/Training',  # This is the source directory for training images
        target_size=(224, 224),
        batch_size=60,
        # Since we use binary_crossentropy loss, we need binary labels
        class_mode='binary')

# Flow training images in batches of 128 using train_datagen generator
validation_generator = validation_datagen.flow_from_directory(
        '/content/drive/My Drive/Colab Notebooks/Validation',  # This is the source directory for training images
        target_size=(224, 224),
        batch_size=10,
        # Since we use binary_crossentropy loss, we need binary labels
        class_mode='binary')

Found 600 images belonging to 2 classes.
Found 100 images belonging to 2 classes.
```

Fig. 6 Compilation process of the proposed model

```
r = model.fit(
        train_generator,
        steps_per_epoch=10,
        epochs=20,
        verbose=1,
        validation_data = validation_generator,
        validation_steps=10)
```

Fig. 7 Training the model

disease, lung disease, stroke or reduced immunity, symptoms progressed, high blood pressure, kidney disease, change in appetite, and loss of sense of smell and one target variable—'Corona Result'. It has values as 0,1 and 2. The number denotes the severity of the symptoms of the patients and denotes them as mild, critical, and severe, respectively. As the data is structured and continuous, it saves us from doing any pre-processing and normalization of the data (Abdel-Basset et al., 2021b; Alankar et al., 2021; Chang et al., 2021a, b; Rojas et al., 2008; Tai et al., 2021).

Feature selection helps in tackling the high features to samples ratio of the dataset. The filter method takes the criteria of relevance score and threshold. These methods evaluate each feature without depending on the prediction method to be used next. The features are processed in a univariate manner, resulting in the classifier's poor performance because feature dependencies can be ignored due to it Hancer et al. (2018). Features with low relevance scores are removed in this method. Various multivariate filter methods are proposed to deal with this disadvantage of the filter approach (Brumpton and Ferreira, 2016; Rouhi, A., & Hossein, 2018; Zhou et al., 2017). Common Filter methods used are Relief, Mutual Information, Joint Mutual Information, fisher Score, Pearson Correlation, and Model-based Ranking (Peng et al., 2005; Duda et al., 2000; Robnik-Šikonja & Kononenko, 2003). Filter models don't depend on the classifier because they select features without any information of classifier to be used in further steps. So, they are fast and straightforward in computing (Ang et al., 2015). Wrapper models are more accurate than Filter models, but they are computationally expensive because they use an embedded classifier to gauge their performance. So, they are dependent on the learning algorithm. For example, for gene expression data, the combinations of genes can be used in better understanding of cancer type and will the possibility to design the experiments for new outputs.

Feature Selection: All the variables of the dataset may not be the features for the target variable or show a correlation with the target variable. However, we can find which one does by calculating the Pearson Correlation between the variables and can filter out the one that shows a positive correlation with the target variables. After performing the following task, we are left with ten variables, including our target variable. We can filter out the data accordingly, as shown in Fig. 8.

(2) *Building the Classification Model:* Before starting with building the model, we must split the data into train and test sets to train the model and test the built model, respectively. So, we must divide the data into dependent.

And independent variables, followed by splitting the data using the train test split() function of learning. The proposed model uses 80:20 ratios for training and testing, respectively. At last, we get the shape of the result, as shown in Fig. 9.

Now, we can build a classification model on the following data, but to make it easier for the machine learning model to learn from the data, we must scale our data and that can be done using Standard Scaler. Later, we define a Logistic Regression model and fit the model with our training data, as shown in Figure 10.

```
data = df[['breathing problem', 'drowsiness', 'pain in chest', 'travel history to infected countries', 'Loss of sense of smell',
        'Corona result', 'body temperature', 'Dry Cough', 'symptoms progressed', 'diabetes']]
```

Fig. 8 Feature Extraction

```
Shape of x:   (127, 9)   | Shape of x_train:  (101, 9)
Shape of y:   (127,)     | Shape of x_test:  (26, 9)
                         | Shape of y_train:  (101,)
                         | Shape of y_test:  (26,)
```

Fig. 9 The shape of the train and test set

```
scaler = StandardScaler()
# Fit on training set only.
scaler.fit(x_train)
# Apply transform to both the training set and the test set.
x_train = scaler.transform(x_train)
x_test = scaler.transform(x_test)

model = LogisticRegression(solver = 'lbfgs')
model.fit(x_train, y_train)
```

Fig. 10 Logistic Regression model

3 Experimental Results

A. Proposed Convolution Neural Network (CNN) Model

The confusion matrix obtained after executing the proposed CNN model is shown in Table 1. For our model, we have noted average Precision (P), Recall (R), and f1-score. Our proposed model classified normal condition vs. COVID-19 condition. The results provided in Fig. 11 shows averages over 20 test cycles.

As we used an image generator and ran our model for 20 epochs, we get the accuracy and loss value of the training generator and validation generator as follows:

i. Training accuracy—0.8850
ii. Training loss—0.2727
iii. Validation accuracy—0.7700
iv. Validation loss—0.4982

We then plotted the results to show the Training and Validation accuracy and loss as shown in Fig. 12.

Table 1. Confusion matrix

	0	0
0	28	21

```
              precision    recall  f1-score   support

           0       0.64      0.68      0.68        50
           1       0.64      0.62      0.62        50

    accuracy                           0.64       100
   macro avg       0.64      0.64      0.64       100
weighted avg       0.64      0.64      0.64       100
```

Fig. 11 Classification Report

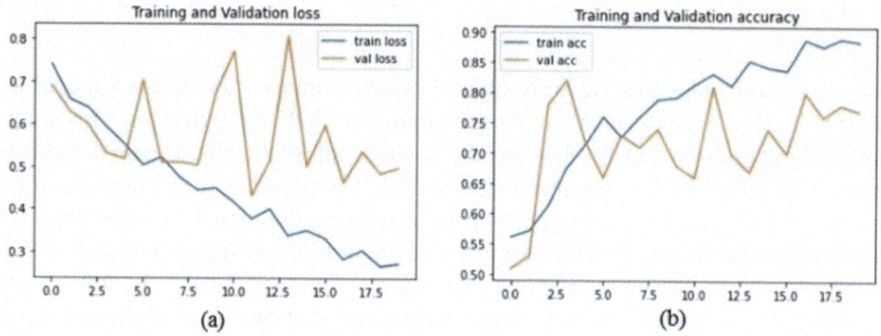

Fig. 12 Visualization of **a** Training and validation loss, **b** Training and validation accuracy

Table 2 Confusion matrix for multi-class data

	0	1	2
0	4	3	0
1	0	15	1
2	0	1	2

B. Patient Sorting Classification Model

After the model is built, a prediction is made on the test set of the data and we can see if the model has made correct predictions or not, using the confusion matrix as shown in Table 2.

The experimental results are listed in Fig. 13 and it includes some evaluation metrics such as average Precision (P), Recall (R), and f1-score. This result is from the classification report of the model. We get an accuracy of 81% on the built model.

	precision	recall	f1-score	support
0	1.00	0.57	0.73	7
1	0.79	0.94	0.86	16
2	0.67	0.67	0.67	3
accuracy			0.81	26
macro avg	0.82	0.73	0.75	26
weighted avg	0.83	0.81	0.80	26

Fig. 13 Classification report

4 Conclusion

Our study concludes that we were able to build an image classification model of chest CT scan images, which can be used to detect COVID-19 disease. This study must be kept updated with the new resources and details of the report studied. Several factors can affect the performance of the model. The accuracy of the model can be increased with the use of a larger data set or training the model for more than 20 numbers of epochs and by adding more convolution layers in the model or using different filters for the input image. A better or efficient model can be built if we get a good quality corpus of CT scan images in the future that can be fast in the diagnosis of COVID-19. As the COVID-19 disease progress throughout the world, an ability to diagnose COVID-19 from a CT scan image would prove helpful in slowing down the spread of the disease and help the medical management system. The Multi-class Patient Sorting classification model showed a good result with the testing data, and the results can be better if the model is trained on bigger data. The patients can be sorted accordingly, and respective measures can be taken accordingly. The mild category patients can be taken to COVID-19 care centers like lodges, stadiums, schools, hotels, and hostels. The severe category patients will be either hospitalized or kept in an extended part of hospitals. The critical category patients will undergo treatment in dedicated COVID-19 hospitals.

Appendix

1 Dr. Harleen Kaur, Post Doc. Research Fellow, Ph.D. (CS) (Fellow, IETE) (Visiting Professor)

Dr. Kaur is an Associate Professor and Chief Investigator at the School of Engineering Sciences and Technology at Jamia Hamdard, New Delhi, India. She is a visiting Professor at TWAS, Italy. She recently worked as Research Fellow at United Nations University (UNU) Tokyo, Japan, International Centre for Excellence-IIGH, Malaysia to conduct research on funded projects from Southeast Asian Nations (SEAN). She is currently working as Principal Investigator on an Indo-Poland bilateral International project funded by the Ministry of Science and Technology, India, and the Ministry of Polish, Poland. Recently, Dr. Kaur got a research funding project from the Ministry of Electronics and Information Technology (MeitY) (Govt. of India), India on the Cybersecurity and Internet of Technologies. She has published more than 100 publications in SCI, referred Journals, and esteemed Conferences. She has published three patents and author of several copyrights. She is the author and editor of many books and Journals in the area of Sustainable Development, Machine Learning, and its applications.

She is a member of several international bodies and received many national and international awards in the research area. Her key research areas include information analytics, applied machine learning, and predictive modeling.

Her publications:

https://scholar.google.com/citations?hl=en&user=NcnUvt0AAAAJ&view_op=list_works&sortby=pubdate

2 Iftikhar Alam

Iftikhar Alam is an undergraduate student of Engineering in Computer Science. His area of interest is the application of machine learning algorithms on pandemic medical datasets.

3 Dr. Ritu Chauhan Ph.D. (CS)

Ritu Chauhan has completed her Ph.D. in Computer Science from Jamia Hamdard, University on Applications of Data Mining techniques in Spatial Databases. She graduated from the University of Delhi, New Delhi. She has previously served as a Lecturer in Computer Science, IMI. Currently, she is working as Assistant Professor in Centre for Computational Biology and Bioinformatics, Amity University. She has published numerous research articles in refereed international journals, conference proceedings, and chapters in an edited book. She is a member of several international bodies. Dr. Ritu has received a funded project from the Ministry of Electronics and Information Technology (MeitY) on the Internet of Technologies. Ritu Chauhan received her B.Sc. degree in Maths from the University of Delhi, New Delhi, and her M.Sc. in Computer Science from Jamia Hamdard University, 2004. Further, she has completed her Ph.D. in Computer Science from Jamia Hamdard, University on Applications of Data Mining techniques in Spatial Databases. She has published numerous research articles in refereed international journals and conference proceedings with chapters in an edited book. She has published three patents and author of several copyrights. She has also served as a reputed member of several national and international bodies. Dr. Ritu has received several national and international grants, awards, and fellowships for excellence in the research area. Her research interests include Data Sciences, IoT, Artificial Intelligence, Spatial systems, and Data analysis for varied application domains.

Her publications:

https://scholar.google.co.in/citations?user=tocwM4MAAAAJ&hl=en

4 Dr. Bhavya Alankar Ph.D. (CSE) M.Tech. (CSE) (Fellow, IETE)

Dr. Alankar is currently working as a senior faculty at the Department of Computer Science and Engg. at Jamia Hamdard, New Delhi, India. Previously employed at the National Institute of Technology (NIT), Jalandhar, India. He has done his Ph.D. in Reconfigurable Computing from Uttarakhand Technical University, India, and Masters in Technology in VLSI design from CDAC, Mohali, India and. He has 15 years of teaching and research experience. His research interests are in VLSI design, Cloud computing, Deep learning, Reconfigurable computing. He is the author and editor of books in the area on VLSI, Machine Learning, and IoT. Dr. Alankar has published four patents and author of several copyrights. He has received many awards and recognition from International bodies.

His publications: https://www.researchgate.net/profile/Bhavya_Alankar

References

Abbas, A., Abdelsamea, M. M., & Gaber, M. M. (2020). Classification of COVID-19 in chest X-ray images using DeTraC deep convolutional neural network. arXiv:2003.13815.

Abdel-Basset, M., Chang, V., & Mohamed, R. (2020). HSMA_WOA: A hybrid novel Slime mould algorithm with a whale optimization algorithm for tackling the image segmentation problem of chest X-ray images. *Applied Soft Computing, 95*, 106642.

Abdel-Basset, M., Chang, V., Hawash, H., Chakrabortty, R. K., & Ryan, M. (2021a). FSS-2019-now: A deep learning architecture for semi-supervised few-shot segmentation of COVID-19 infection. *Knowledge-Based Systems, 212*, 106647.

Abdel-Basset, M., Chang, V., & Nabeeh, N. A. (2021b). An intelligent framework using disruptive technologies for COVID-19 analysis. *Technological Forecasting and Social Change, 163*, 120431.

Alankar, B., Kaur, H., Ahsaan, S., Sharma, G., & Chang, V. (2021). Towards reviewing an immediate impact of COVID-19 on the integrative world economy: An evolving perspective. *Journal of Global Information Management (JGIM), 30*(4), 1–19.

Ang, J. C., Haron, H., & Hamed, H .N. A. (2015). Semi-supervised SVM-based feature selection for cancer classification using microarray gene expression data. In M. Ali, Y. S. Kwon, C.-H. Lee, J. Kim, & Y. Kim (eds.), *Current approaches in applied artificial intelligence* (Vol. 9101, pp. 468–77). Lecture Notes in Computer Science. Springer International Publishing.

Apostolopoulos, I. D., & Mpesiana, T. A. (2020). COVID-19: automatic detection from X-ray images utilizing transfer learning with convolutional neural networks. *Physical and Engineering Sciences in Medicine*, 1.

Brumpton, B. M., & Ferreira, M. A. R. (2016). Multivariate EQTL mapping uncovers functional variation on the X-chromosome associated with complex disease traits. *Human Genetics, 135*(7), 827–839. https://doi.org/10.1007/s00439-016-1674-6

Chang, V., Abdel-Basset, M., Iqbal, R., & Wills, G. (2021a). IEEE transactions on industrial informatics EDITORIAL: Advanced deep learning techniques for COVID-19. *IEEE Transactions on Industrial Informatics, 17*(9), 6476–6479.

Chang, V., Goble, C., Ramachandran, M., Deborah, L. J., & Behringer, R. (2021b). Editorial on machine learning, AI and Big data methods and findings for COVID-19. *Information Systems Frontiers*, 1–5.

Charles, P. W. D. (2013). Project Title, GitHub repository. Retireved from https://github.com/charlespwd/project-title.

Chauhan, R., Kaur, H., & Chang, V. (2021). An optimized integrated framework of big data analytics managing security and privacy in healthcare data. *Wireless Personal Communications, 117*, 87–108. https://doi.org/10.1007/s11277-020-07040-8

Dhiman, G., Chang, V., Kant Singh, K., & Shankar, A. (2021). Adopt: automatic deep learning and optimization-based approach for detection of novel coronavirus covid-19 disease using x-ray images. *Journal of Biomolecular Structure and Dynamics*, 1–13.

Duda, R. O., Hart, P. E., & Stork, D. G. (2000). *Pattern Classification* (2nd ed.).

Foster, K. R., Koprowski, R., & Skufca, J. D. (2014). Machine learning, medical diagnosis, and biomedical engineering research-commentary. *Biomedical Engineering Online, 13*, 94.

Ge'ron, A. (2019). Hands-on machine learning with Scikit-Learn, Keras, and TensorFlow: Concepts, tools, and techniques to build intelligent systems. O'Reilly Media.

Greenspan, H., Van Ginneken, B., & Summers, R. M. (2016). Guest editorial deep learning in medical imaging: Overview and future promise of an exciting new technique. *IEEE Transactions on Medical Imaging, 35*(5), 1153–1159.

Gulsen, A. (2020). A simple classification of COVID-19 patients. *7*, 62–63. https://doi.org/10.15406/jlprr.2020.07.00230.

Han, J., Qian, K., Song, M., Yang, Z., Ren, Z., Liu, S., et al. (2020). An early study on intelligent analysis of speech under COVID-19: Severity, sleep quality, fatigue, and anxiety. arXiv:2005.00096.

Hancer, E., Xue, B., & Zhang, M. (2018). Differential eEvolution for filter feature selection based on information theory and feature ranking. *Knowledge-Based Systems, 140*, 103–119. https://doi.org/10.1016/j.knosys.2017.10.028

https://www.kaggle.com/allen-institute-for-ai/CORD-19-research-challenge.

Kaur, H., Wasan, S. K. (2006). Empirical study on applications of data mining techniques in healthcare. *Journal of Computer Science, 2*(2), 194-200.

Kaur, H., Ahsaan, S. U., Alankar, B., et al. (2021). A proposed sentiment analysis deep learning algorithm for analyzing COVID-19 tweets. *Information Systems Frontiers*. https://doi.org/10.1007/s10796-021-10135-7

Khan, I. U., & Aslam, N. (2020). A deep-learning-based framework for automated diagnosis of COVID-19 using X-ray images. *Information, 11*(9), 419.

Lee, C. M., & Ko, C. N. (2009). Time series prediction using RBF neural networks with a nonlinear time-varying evolution PSO algorithm. *Neurocomputing, 73*(1–3), 449–460.

Lu, X., Wang, Y., Chen, T., Li, Y., Wang, J., & Yan, F. (2020). Classification of COVID-19 in intensive care patients: towards rational and effective clinical triage. medRxiv.

Menezes Jr, J. M. P., & Barreto, G. A. (2008). Long-term time series prediction with the NARX network: An empirical evaluation. *Neurocomputing, 71*(16–18), 3335–3343.

Narin, A., Kaya, C., & Pamuk, Z. (2020). Automatic detection of coronavirus disease (COVID-19) using x-ray images and deep convolutional neural networks. arXiv:2003.10849.

Peng, H., Long, F., & Ding, C. (2005). Feature selection based on mutual information: Criteria of max-dependency, max-relevance, and min-redundancy. *IEEE Transactions on Pattern Analysis and Machine Intelligence, 27*(8), 1226–1238. https://doi.org/10.1109/TPAMI.2005.159

Robnik-Šikonja, M., & Kononenko, I. (2003). Theoretical and empirical analysis of ReliefF and RReliefF. *Machine Learning, 53*(1/2), 23–69. https://doi.org/10.1023/A:1025667309714

Rojas, I., Valenzuela, O., Rojas, F., Guillén, A., Herrera, L. J., Pomares, H., & Pasadas, M. (2008). Soft-computing techniques and ARMA model for time series prediction. *Neurocomputing, 71*(4–6), 5191–537.

Rouhi, A., & Hossein, N. (2018). Filter-based feature selection for microarray data using improved binary gravitational search algorithm. In *2018 3rd Conference on Swarm Intelligence and Evolutionary Computation (CSIEC)* (pp. 1–6). Bam: IEEE.

Singh, A., Jindal, V., Sandhu, R., & Chang, V. (2021). A scalable framework for smart COVID surveillance in the workplace using Deep Neural Networks and cloud computing. *Expert Systems*, e12704.

Sorjamaa, A., Hao, J., Reyhani, N., Ji, Y., & Lendasse, A. (2007). Methodology for long-term prediction of time series. *Neurocomputing, 70*(16–18), 2861–2869.

Tai, Y., Gao, B., Li, Q., Yu, Z., Zhu, C., & Chang, V. (2021). Trustworthy and intelligent covid-19 diagnostic it through XR and deep learning-based clinic data access. *IEEE Internet of Things Journal, 8*(21), 15965–15976.

Thakur, S., & Kumar, A. (2021 Augest). X-ray and CT-scan-based automated detection and classification of covid-19 using convolutional neural networks (CNN). *Biomedical Signal Processing and Control, 69*, 102920.

Wang, L. (2020). A tailored deep convolutional neural network design for detection of COVID-19 cases from chest X-ray images. arXiv:2003.09871.

Yang, H., Huang, K., King, I., & Lyu, M. R. (2009). Localized support vector regression for time series prediction. *Neurocomputing, 72*(10–12), 2659–2669.

Zhang, J., Xie, Y., Li, Y., Shen, C., & Xia, Y. (2020). COVID-19 screening on chest X-ray images using deep learning-based anomaly detection. arXiv:2003.12338.

Zhou, Y., Pei, W., Xianlong, W., Ji, Z., Peter, X. -K. S. (2017). Sparse multivariate factor analysis regression models and its applications to integrative genomics analysis. *Genetic Epidemiology, 41*(1), 70–80. https://doi.org/10.1002/gepi.22018.

Coordination of Covid-19 Vaccation: An Optimization Problem and Related Tools Derived from Telecommunications Systems

Christophe Gaie and Markus Mueck

Abstract The discovery of vaccines against Covid-19 is coupled with an initial shortage due to the time required to produce them and vaccinate people. This raises the crucial issue of vaccine allocation to optimize people protection, especially the most vulnerable, as they are more likely to contract a severe form of the disease. First, this chapter proposes a review of the existing literature on prioritization criteria to consider as well as methods to optimize the allocation of vaccines and avoid deaths or long-term impacts of the disease. Then, the authors propose a novel decentralized approach to optimize vaccine allocation. A novel way is presented for combining traditional convex optimization problems with Artificial Intelligence (AI) tools. The latter tools are used to characterize the boundary condition for the convex optimization problem, which is then used applying Lagrangian optimization methodologies. AI is thus applied to suitably frame the problem instead of directly deriving the final solution—in this approach, the advantages of AI are exploited while maintaining a high level of explainability for the overall optimization approach. Finally, a discussion presents the outcomes expected from this cross-domain approach.

Keywords Data science · Data analytics · Artificial intelligence · Covid-19 · Pandemic · Vaccination

1 Introduction

Two years after the beginning of the Covid-19 pandemic, the sanitary and economic situation worldwide remains highly worrying. The world economy was severely affected by the multiple restrictions with a high impact on the stock exchange (Alankar et al., 2021). Almost every economic sector was hit, but the restrictions

C. Gaie (✉)
Centre Interministériel de Services Informatiques RH, Paris, France
e-mail: christophe.gaie@gmail.com

M. Mueck
Unterhaching, Germany

V. Chang et al. (eds.), *Artificial Intelligence and Machine Learning Methods in COVID-19 and Related Health Diseases*, Studies in Computational Intelligence 1023,
https://doi.org/10.1007/978-3-031-04597-4_11

were harder for tourism, education, entertainment, or sports. Thankfully, policy-makers have taken multiple measures to protect the population and the economy, including monetary and fiscal measures, healthcare and human protection decisions, and digital business incentives.

However, the virus is still threatening, companies struggle with cash-flow difficulties or low profitability, and above all, people are tired of observing multiple constraints that reduce contamination and social life. The situation was emphasized by social media network platforms (e.g., Facebook, Twitter) that convey worrying as well as false information (Kaur et al., 2021). This largely impacts people's mental health status and may contribute to the emergence of psychic disorders.

Multiple innovations enable the population to face the pandemic. For instance, the introduction of intelligent diagnostic by combining data, image, and computing (Tai et al., 2021) offers outstanding perspectives to fight against the pandemic. Another promising approach detects potentially infected persons through video detection (Singh et al., 2021). This novel system offers very high accuracy and reduces propagation risks as contamination chains may be easily broken. In both proposals, the usage of cloud computing facilitates and accelerates the deployment of the innovation, which is a key concern in the context of the race against time.

Another interesting approach aims to overcome the Covid-19 test kit shortage by using X-ray chest images (Dhiman et al., 2021). The proposal combines optimization and deep learning-based techniques to identify Covid-19 contaminated patients. Several of the convolutional neural network-based (CNN) models tested show a particularly high accuracy (more than 98%). This demonstrates the interest in using X-ray detection of Covid-19 in the context of test kit shortage.

Above the previous innovations, the emergence of vaccination capabilities is an opportunity to overcome difficulties. Governments have to organize vaccination campaigns and optimize their efficiency to avoid the multiplication of new contamination waves. That is why proven methods or technology constitute an essential asset in the time trial against the pandemic.

Optimizing the organization of a vaccination campaign requires injecting an active substance into patients and tackling a complete process. This begins with the fabrication process of vaccines (which is considered beyond the scope of this research). Then, the optimization requires collecting a large amount of data to define who will benefit from vaccines and when and where the injection will be performed. This computation raises multiple concerns regarding data protection, prioritization strategy, transparency, etc. (Fig. 1).

In the upper context, the following specific problem statements are addressed in the present paper:

1. Formalization of the vaccination strategy as a convex optimization problem. Specific numerical optimizations objectives are defined (including the number of deaths, etc.) as well as boundary conditions (e.g., the available capacity of hospitals, etc.); furthermore, context information is injected into the problem statement (e.g., the current level of Covid-19 infections, vaccination status, etc.).

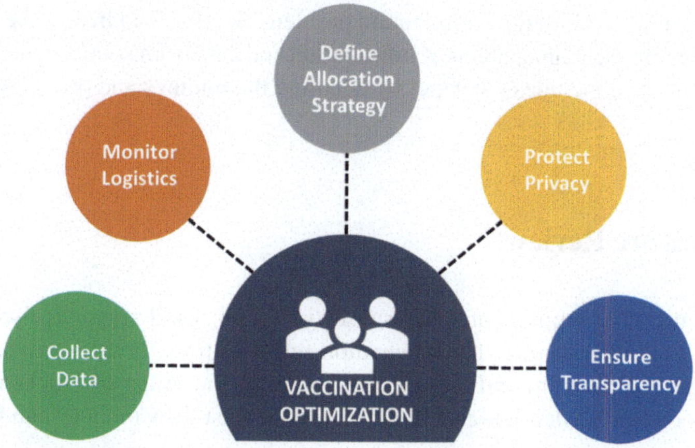

Fig. 1 Illustration of the key concerns to optimize vaccination

2. A novel way is presented on combining traditional convex optimization problems with Artificial Intelligence (AI) methods. AI proves indeed very useful to process a large number of observations related to Covid-19, such as data on local penetration of infections, local vaccination status, etc. In principle, AI techniques may be applied to solve the optimization problem as a whole. However, the drawback often lies in the limited explainability of the inherent solution; i.e., the AI components (such as deep neural networks and similar) are often seen as black boxes that "somehow" derive meaningful results, but the inherent reasoning is hard to decipher. In our approach, AI is refining the boundary conditions and context/status information, but the actual optimization is executed building on well-understood Lagrangian methods. This finally leads to a high level of trust in the final result of the process.
3. To meaningfully react in the context of the Covid-19 pandemic situation, a quick reaction to an ever-changing situation is required—in particular, on a local level, a seemingly stable situation may go out of control within a short time. Reliable information is a key resource to support the decision-making process. A key challenge lies in the vast amount of unreliable and often false information or propaganda. Originally correct information may furthermore be misrepresented and falsified to support certain views. It is therefore essential to share information so that the integrity of the original message is maintained. The authorship is documented so that it cannot be modified and new information is added in a well-organized and verifiable way. The authors demonstrate how the novel Distributed Ledger Technology paradigm can achieve those objectives.

The chapter is organized as follows. Section 2 provides a literature review of the usage of technology to fight against the pandemic and optimize vaccination. Section 3 formalizes the convex optimization problem such that optimum vaccination strategies may be derived. Section 4 further supports the approaches detailed in the previous Section by using AI to refine the boundary conditions and to update context information injected into the optimization problem. Section 5 addresses the problem of how to secure the management of information and data from source to destination. Section 6 finally concludes the paper by detailing the findings and providing further perspectives.

2 Literature Review

The literature is composed of three subsections that detail respectively: the risk criteria to take into account for vaccination prioritization, the actions required to organize the vaccination, and a brief description of the usage of blockchains and Permissioned Distributed Ledger (PDL) to fight against the Covid-19 pandemic.

It is of the highest importance to identify risk parameters to consider to define a vaccination priority policy.

In Gong et al. (2020), propose a method to predict the apparition of a severe form of Covid-19 among contaminated patients. The approach relies on seven features, including the age of patients as well as a focus on immune cells. Whereas this approach would be really interesting to identify the population to vaccinate, it requires a post-contamination analysis. This prevents governments from defining a preventive vaccination strategy.

In Leng et al. (2020), propose very informative research concerning individuals' decision-making concerning Covid-19 vaccination. Their research concerns six Chinese provinces and relies on seven criteria: vaccine effectiveness, side-effects, accessibility, number of doses, vaccination sites, duration of vaccine protection, and proportion of acquaintances vaccinated. They prove the importance of each of these criteria, but the method relies on their hypotheses which are verified through polling.

In Estiri et al. (2021), detail a method to predict the probability of death after contracting the Covid-19 disease. The approach uses the patient's medical history and comorbidities to build a prediction model. The results are relevant and offer similar results compared to methods requiring symptoms, laboratory values, and images at the time of diagnosis. The research identifies high-risk populations such as patients at ages of 45 and 65 years with diabetes or cancers as well patients aged 65–85 years with pulmonary diseases or smoking history. Although the approach is relevant to prioritizing vaccination, the results are fairly predictable.

In Schwarzinger et al. (2021), tackled the topic of COVID-19 vaccine hesitancy. Their approach relies on a survey experiment including around two thousand adults aged 18 to 64 years old grouped by gender, age, education, household size, and residence. The responses indicate that 28·8% of the sample are categorically opposed to vaccination. Interestingly, the research underlines that the opposition to the vaccine is significantly associated with female gender, age (with an inverted U-shaped relationship), lower educational level, poor compliance with recommended vaccinations in the past, and no report of specified chronic conditions (i.e., no hypertension). These findings should play an important role in anticipating citizen behavior and optimize the allocation of vaccines.

In Matrajt et al. (2021) describe a vaccine allocation system that relies on an age-stratified model and optimization algorithms combining four metrics (deaths, symptomatic infections, and maximum hospitalizations in intensive or non-intense care units). Interestingly, the optimal vaccine allocation depends on the vaccine's effectiveness. For instance, low vaccine effectiveness requires prioritizing high-risk age groups to minimize deaths. On the opposite, if the vaccine effectiveness is high, the best allocation consists in vaccinating the high-transmission age groups. While this research is really promising, it does not offer a direct implementation as it is based on a centralized approach with a single vaccine option.

Above the literature concerning risk identification, there is a huge interest of researchers towards proposing actions to organize the vaccination.

In Abo et al. (2020), Abo and Smith describe the effect of vaccination using the risk equation. The study outlines two parameters with the highest impact on the pandemic: the transmission probability and the fraction of people using the vaccine. The research indicates that the situation may be worst if vaccination is lower than 80% and protection protocols are abandoned. Whereas the analysis interestingly underlines the importance of seeking a high vaccination coverage of the population, the interest in vaccination does not seem questionable and the termination of protection protocols seems very improbable.

In Mcdonnell et al. (2020), indicate that designing a vaccine does not constitute a magic solution against the Covid-19 pandemic. The paper outlines the necessity to follow along and difficult process from the finding of a vaccine to its validation by sanitary authorities provided it offers sufficient protection against the virus. They propose to develop multiple solutions to fight against the virus such as investing in multiple vaccine candidates through international collaboration and market.

Incentives. Today, this proposal reveals appropriate as multiple variants of Covid-19 appeared and some vaccines reveal more efficiency than others. Whereas this idea is useful, the paper does not indicate how to benefit from this diversity of measures.

In Matrajt et al. (2020), outline the importance to prepare for vaccine scarcity as soon as the medicine is available. The authors propose a vaccine allocation algorithm to solve an optimization problem composed of four metrics (deaths, symptomatic infections, and maximum non-ICU and ICU hospitalizations). Interestingly the authors prove that if the produced vaccines show low effectiveness then the best choice is to protect high-risk age groups first. On the contrary, for higher vaccine effectiveness, a counter-intuitive approach should be followed and would lead to allocating vaccines to high-transmission age groups. The research also tackles the initial immunity of the population as well as the progressive arrival of vaccines doses. The main difference with the approach proposed in the current chapter is the centralized approach at a State level compared to the distributed method of this chapter.

In Kim et al. (2021), describe the different subjects to consider to coordinate a massive vaccination campaign. The authors underline 4 successive steps to overcome: obtain vaccine approval, manufacture vaccine doses, operationally organize the vaccination and define optimization measures to prepare the future campaign. This paper underlines the complexity of the vaccination campaign and outlines some useful aspects. However, it does not propose a structured approach to fight against the pandemic.

In Tang et al. (2021), analyze two opposite effects: the benefits of immunization obtained through vaccination on the one side; and the drawbacks of reducing sanitary constraints on the population as a vaccination counterpart. The authors demonstrate that the positive effects depend on the vaccine effectiveness as well as the evolution of the virus transmissibility. This advocates for an approach combining and monitoring every protection measure (vaccination, protection gesture) to adjust the rapidly sanitary constraints.

In Bubar et al. (2021), proposed a mathematical model to analyze the efficiency of five age-stratified vaccination strategies. The research proves that reducing the transmissibility of the virus requires prioritizing adults aged 20 to 49 when a highly effective transmission-blocking vaccine is available. On the opposite, the vaccination of adults greater than 60 years old enables to reduce mortality and maximize the number of years lived. The authors underline the evolutivity of their framework which may be adjusted for the younger generation (under 18 years old) and the elder (above 70 or 80 years old). As a matter of fact, the findings of their work may be reused in the largest context of allocating vaccine doses to the population according to a specific objective.

In Han et al. (2021), describe a vaccine allocation algorithm that introduces dynamic adaptation to the pandemic. This approach takes advantage of the continuous evolution of the situation and introduces up-to-date information to adjust the vaccination strategy. The authors underline the benefits of simultaneously reducing the number of deaths and infections to the virus. In particular, the vaccination effort should be increased as soon as a new wave appears. However, the research presents an important limit since people under 20 and over 60 are excluded from the vaccine optimization.

In Buhat et al. (2021), develop an algorithm to optimize the allocation of COVID-19 vaccines in the Philippines to minimize the number of deaths while protecting the vulnerable population. The proposal relies on a linear programming model that takes into account multiple parameters: the governmental vaccine prioritization strategy, the vaccine supply, efficiency, and cost, as well as the projection of deaths. This model offers a useful framework to assist policymakers when they define a vaccination strategy.

In the remainder of this literature review, a brief description of the usage of blockchains and PDL to fight against the Covid-19 pandemic is proposed.

In Rauchs et al. (2018), propose a conceptual framework concerning Distributed Ledger Technology. The architecture is composed of three interdependent layers: 1- the Protocol layer defines a set of rules that determine how the system functioning; 2- the Network layer that illustrates how the actors are interconnected and what are the processes to implement; and 3- the Data layer that provides information enabling the system functioning. The proposal reveals attractiveness to model the Distributed Ledger Technology but does not offer information on its concrete operation implementation.

In Jabarulla and Lee (2019), present a conceptual framework to tackle Covid-19 pandemic chanllenges. The framework is decentralized and relies on a patient-centric approach relying on blockchain and artificial intelligence (AI) technologies. This offers multiple advantages such as interoperability of different stakeholders, security and confidentiality of patients' health folders, secured exchange of information, and the possibility to enforce federated learning using up-to-date AI models. The paper offers a very interesting conceptual approach that is advantageously complemented by operational research.

In Suhasini and Dilbag (2021), describe a technical approach to implement a blockchain-based framework for secure data management of the healthcare system. The authors detail a system based on smart contracts and a unique electronic health record for patient data management. Their proposal details the comprehensive developments performed as well as test cases for validation. This paper confirms the applicability of the blockchain approach and offers details on possible implementation.

In Gupta et al. (2021), propose a using blockchain-based framework that is dedicated to monitoring the COVID-19 testing kit supply chain management. The proposal offers three principal advantages: 1- high security and transparency provided that no unapproved party can be added to the chain; 2- permanency of the information as the blockchain stores data profoundly through cryptographic calculations and validation by a trustworthy third-party; and 3- user verification as the final consumer may verify the approval through the batch number. Whereas the usage of blockchain is attractive it only tackles a very specific in the fight against the Covid-19 pandemic.

In Antal et al. (2021), underline the importance of defining an operational and transparent distribution to ensure the success of the Covid-19 vaccination campaign. To this aim, the authors propose to implement a method that relies on blockchain technology which ensures data integrity and inalterability of vaccination registries. Their approach combines information from beneficiaries, medical centers, practitioners, and vaccine producers. The authors implemented the prototype which also takes into account the COVID-19 vaccine distribution conditions. This implementation was verified on the Ethereum test network, Ropsten, and proved its efficiency in this context.

In Carniel et al. (2021), propose a method to ensure a reliable vaccine allocation through the usage of blockchain technology. The proposal combines information obtained from five stakeholders: the government, the patients, the practitioners, the vaccine centers, and the vaccine manufacturers. The approach proposed is global and involves many actors contributing to a common immunization objective. The research reveals valuable as a prototype is described and relies on Ethereum Blockchain. However, the security aspects of the process are unfortunately excluded from the scope.

In ETSI (2021), the European Telecommunications Standards Institute develop a framework based on Permissioned Distributed Ledger to implement Federated Data

Management. The proposal details multiple use cases and applications of the framework. It also details the key issues concerning data collection, storage, and federation. The architecture described reveals generic which facilitates its implementation in various contexts. These advantages justify that the current paper largely inspires from the content to propose a Covid-19 vaccination framework.

In Arani et al. (2020), develop a framework that takes advantage of Permissioned Distributed Ledger to ensure a distributed and secured data storage and restitution. As a matter of fact, the proposed architecture relies on three layers: an end-user registration, a patient's record right management, and a transaction management module. Whereas this proposal is promising, the paper does not illustrate the concrete functioning of the system through a prototype.

3 Modelization and Introduction of the Vaccine Allocation Algorithm (VAA)

a- **Description of hypotheses, objectives, and constraints**

In this paper, the authors aim to tackle the problem of finding the best allocation of vaccine doses to maximize the number of years lived in a given zone. The model introduces multiple evolutive constraints such as regional fairness, age, and health criteria or liberty impacts.

The resolution of discrete optimization problems was deeply detailed in Boyd and Vandenberghe (2004). The authors already implemented algorithms derived from the Lagrangian theory in Gaie et al. (2008). Further details and explanations were brought in Gaie (2010). These methods require defining both the objective to optimize and constraints to fulfill. Thereafter some possible objectives and constraints are described to define the problem to solve (Fig. 2).

Alternative Objectives:

- **Minimize the number of deaths,** aims to reduce the number of deaths. This requires to define a vaccination strategy that protects vulnerable people as well as reduces the contamination diffusion of the virus.
- **Maximizing the number of years lived,** aims to protect people whose life expectancy is high as with the significant risk of contamination and probability of contracting a severe form of the covid-19 disease. As the number of vaccine doses is scarce, this scenario could require "sacrifice" immunosuppressed persons with low life expectancy.
- **Ensure a tradeoff with the economy,** aims to xxx

Resolving the optimization problem with each objective would provide different solutions. Indeed, minimizing the number of deaths would probably require high protection of the elder whereas ensuring a tradeoff as proposed in (Gaie & Mueck,

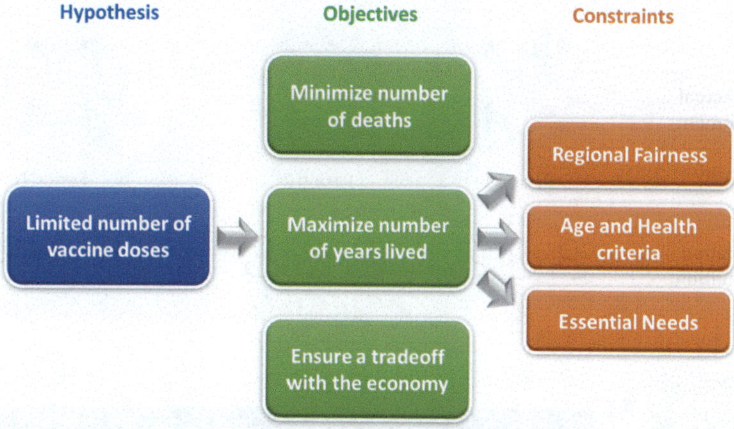

Fig. 2 Vaccination problem schematization

2022) would ensure that young people are protected early as their contribution to the economy is strongly reduced by confinement measures.

In the remainder of this paper, the authors chose to retain the "**Death Minimization objective,**" which reveals to be the more widespread strategy adopted all around Western countries (Europe and United States of America).

Cumulative Constraints:

- **Regional fairness** intends to ensure equality between regions or departments of a country. This means that the number of vaccine doses granted may be proportional to the number of inhabitants. For instance, metropolitan areas such as Paris, Berlin, or Madrid should obtain more vaccine doses than rural areas such as respectively Normandie, Brandebourg, or Castille-et-León.
- **Age and health criteria** intend to ensure that vaccine doses are scattered fairly among the different generations. Although the sanitary measures struck every generation, the young generation made a huge effort by endorsing the economic impacts in the long term to protect the elder. To facilitate the restart of the economy and improve the acceptability of sanitary measures, ensuring fairness in the vaccination process may be useful. The fairness may be perfect or adjusted to provide a reasonable amount of vaccine doses to every generation. For instance, each generation could be granted a minimal amount of 10–20% of vaccine doses.
- **Essential needs** aim to give priority to a person to vaccinate according to national priorities: education, healthcare, security, etc. A critical objective would be to grant more doses to young people and teachers to guarantee that their education will not be disturbed anymore. A similar approach could be employed for the person whose job cannot be performed by teleworking: this corresponds to medical practitioners, supermarket employees, or essential civil servants (firemen, policemen, etc.).

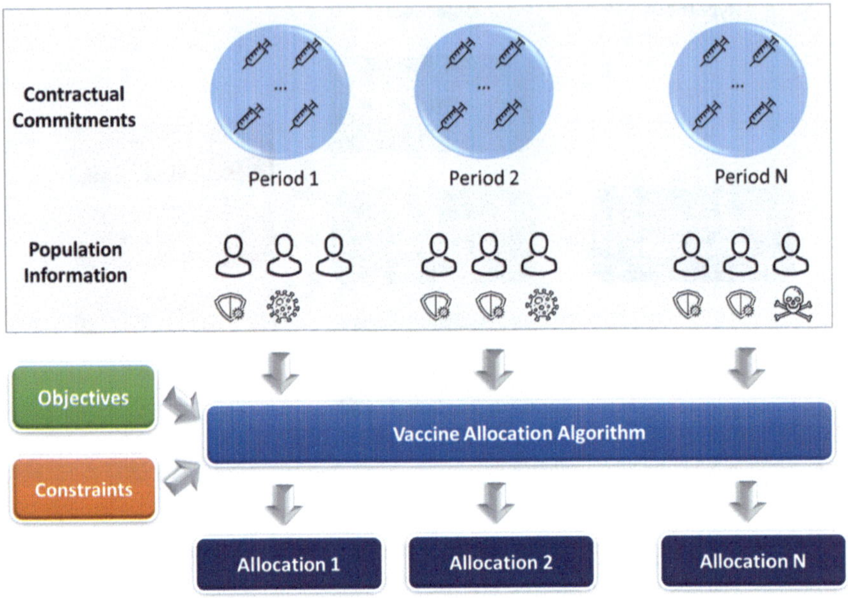

Fig. 3 Illustration of the evolutive allocation using the VAA

In the remainder of this paper, the authors retain the first two constraints and consider that essential needs may be managed by booking specific vaccine doses in a dedicated vaccination line.

b- Introduction of the Vaccine Allocation Algorithm (VAA) (Fig. 3)

First, define the following variables:

- i corresponds to the region considered
- i, j corresponds to the district considered in the region i
- i, j, k corresponds to the town considered, in the district j of region i
- n corresponds to a period

The local parameters are defined as follows:

- $N_{i,j,k}(n)$ the number of inhabitants for the period n in location *i,j,k*
- $\Delta_{i,j,k}(n)$ the number of vaccine doses for the period n in location *i,j,k*

Then, define $s_p(n)$ the sensitivity to develop a severe form of Covid-19 for the person p. This parameter depends on:

- $a_p(n)$ the age of the person
- $v_p^i(n)$ the initial vaccination status of the person in the period
- $\delta_{i,j,k,p}(n)$ the allocation of a vaccine dose in the period n for person p in location *i,j,k*

- $v_p^f(n)$ the final vaccination status of the person in the period

$$s_p(n) = f_1(a_p(n), v_p^i(n), \delta_{i,j,k,p}(n), v_p^f(n))$$

Let also define $m_p(n)$, the mobility of the person and $b_p(n)$, the behavior of a person p with regards to the protection against the virus (respect of social distance, face masks, hand washing, etc.). Then, introduce the variable $r_p(n)$ which corresponds to the risk for a person p to be contaminated through mobility and behavior of herself and others:

$$r_p(n) = f_2(\{m_{p'}(n)\}, \{b_{p'}(n)\})$$

Define $c_{i,j,k,p}(n) = c_{i,j,k}(n) * s_p(n) * r_p(n)$ as the probability that a person p may suffer from a severe form in town k on the period n.

Define $d_{i,j,k,p}(n) = f_3(c_{i,j,k,p}(n), o_p(n), h_p(n))$ as the probability that a person p may die of Covid-19 in town k on the period n. This parameter depends on:

- $o_p(n)$ the occupancy of hospitals and the availability of practitioners and medicine
- $h_p(n)$ the health status of the person, which is completely private

Afterward, define the following functions:

- $D_e(n)$ is the expected number of estimated deaths on the period n (a priori)
- $D_m(n)$ is the number of deaths measured on the period n (a posteriori)

Thus, $D_e(n)$ is computed as follows:

$$D_e(n) = \frac{D_e(n-1)}{D_m(n-1)} \sum_i \sum_j \sum_k \sum_p d_{i,j,k,p}(n)$$

The immediate idea is to solve the following problem (P1) without any constraint:

(P1) Allocate $\delta_{i,j,k,p}(n)$ such as Minimizing $D_e(n) = \frac{D_e(n-1)}{D_m(n-1)} \sum_i \sum_j \sum_k \sum_p d_{i,j,k,p}(n)$

This centralized approach requires obtaining a tremendous amount for each period, relies on a centralized algorithm, and does not ensure any fairness (between regions, generations, etc.). To introduce equality, the authors propose to introduce constraints and solve a new problem. First, we define a class of age C_l as a group of people with a range of ages with a defined sampling step (e.g., five years). Thus $\Delta_{i,j,l}(n) = \sum_k \sum_p \delta_{i,j,k,p,l}(n)$ corresponds to the number of doses granted to this generation in department j. Then, the ratio of vaccinated persons among a class of age may be expressed as $\gamma_{i,j,\,l}(n)$. Define (P2) as the following optimization problem:

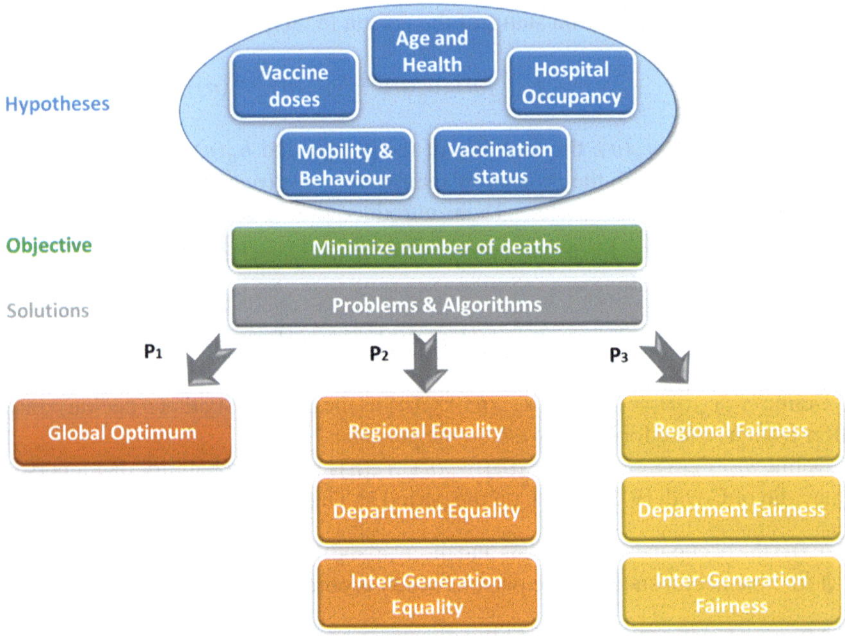

Fig. 4 Illustration of the algorithm functioning

(P2) Allocate $\delta_{i,j,k,p,l}(n)$ such as Minimizing $D_e(n) = \frac{D_e(n-1)}{D_m(n-1)} \sum_i \sum_j \sum_k \sum_p d_{i,j,k,p,l}(n)$ where:

- For each region *i*, $\Delta_i(n) = \frac{N_i(n)}{\sum_i N_i(n)}$ *(regional equality)*
- For each department *j*, $\Delta_{i,j}(n) = \frac{N_{i,j}(n)}{\sum_j N_{i,j}(n)}$ *(department equality)*
- For each class of age C_l, $\Delta_{i,j,l}(n) = \frac{N_{i,j,l}(n)}{\sum_l N_{i,j,l}(n)}$ *inter-generation equality within region i and department j)*

Let now introduce Fairness parameters for each layer (F_i -$F_{i,j}$- $F_{i,j,l}$) whose value ranges from 0 to 1.[1] Define (P3) as the following optimization problem (Fig. 4):

(P3) Allocate $\delta_{i,j,k,p,l}(n)$ such as Minimizing $D_e(n) = \frac{D_e(n-1)}{D_m(n-1)} \sum_i \sum_j \sum_k \sum_p d_{i,j,k,p,l}(n)$where:

[1] Note that the fairness Distributed Vaccine Allocation Algorithm (DVAA).could also evolve according to the time period *n*.

- For each region i, $\Delta_i(n) \geq F_i \frac{N_i(n)}{\sum_i N_i(n)}$ *(regional fairness)*
- For each department j, $\Delta_{i,j}(n) \geq F_{i,j} \frac{N_{i,j}(n)}{\sum_j N_{i,j}(n)}$ *(department fairness)*
- For each class of age C_l, $\Delta_{i,j,l}(n) \geq F_{i,j,l} \frac{N_{i,j,l}(n)}{\sum_l N_{i,j,l}(n)}$ *(inter-generation fairness within region i and department j)*

4 Artificial Intelligence Solutions Derived from Telecommunications Systems to Support the Optimization Process

The upper optimization process may only lead to suitable vaccination strategies when the related boundary conditions and injected context information is correct, such as the occupancy rate of hospitals, the health characteristics of a local population, etc. In other words, the quality of the optimization results depends on the quality of the input and context parameters.

Often, a plethora of related information is available, however, through large amounts of raw (unprocessed) observations, which are difficult to use without thorough and complex processing. AI is a suitable tool for processing such large amounts of seemingly chaotic information to derive suitable data sets we propose using as parameters for the upper convex optimization problem. Another key advantage to this proposed approach is that AI decision-making typically suffers from poor explainability, i.e., the produced results are often meaningful. However, the derivation process is difficult (impossible) to understand or verify. As a result, there may be an issue of trust, which is, in particular, relevant in such a critical context of national interest as the coordination of a Covid-19 vaccination campaign. By applying the outputs of the AI processing to our convex optimization approach, we achieve to (i) exploit the capabilities of AI to support the modeling of the problem statement while (ii) maintaining a high level of explainability of the approach. Convex optimization and related Lagragian methodologies are well understood and thus achieve the highest levels of trustworthiness.

For the AI process and derivation of boundary conditions and context information injected into the convex problem statement, the authors propose to build on the ETSI Architecture as introduced by ETSI ENI (ETSI Group Specification (GS) 2019) is illustrated in Fig. 5. Further details are provided in the following. The relationship of the various components to a pandemic situation is indicated through examples in Fig. 4.

The first step to be addressed is providing information on the proposed architectural framework. Data may indeed originate from different sources and must be processed to be represented using a well-defined format. The various building blocks of the architecture will then be able to use and process the data. In the context of a pandemic, "Data" may relate to observations on the spreading of the disease over

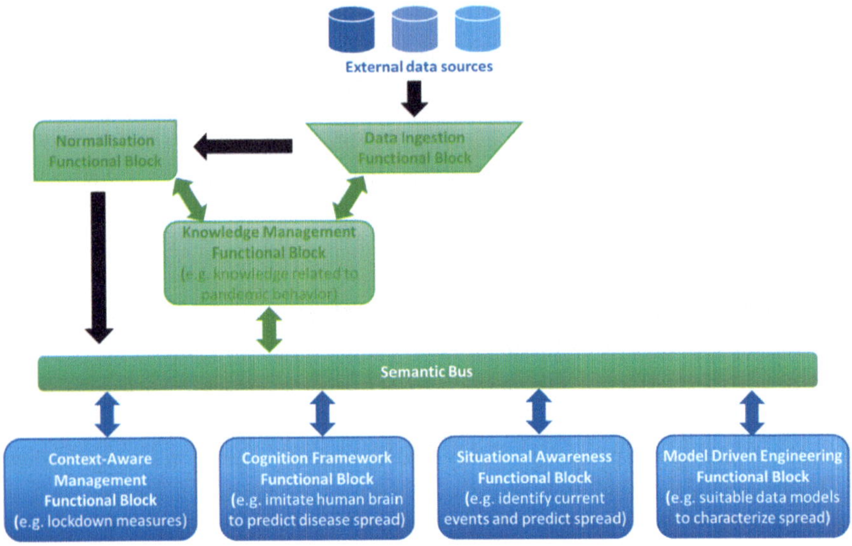

Fig. 5 ETSI Architecture for Artificial Intelligence in Telecommunication Networks

time, characteristics of the population (age, population density, health information, etc.).

This process is typically referred to as "Data Ingestion" and may include the following steps:

i. Data Filtering for removal of unnecessary or not useful information;
ii. Data Correlation to create an association or relationship between data;
iii. Data Cleansing for detection and removal of corrupt, incomplete, inaccurate, and/or irrelevant data;
iv. Data Anonymization and Pseudonymization for removing or protecting (encrypting) data which may be used to identify people;
v. Data Augmentation for adding other types of data to the existing data set to enrich it;
vi. Data Labelling to add class labels to data sets.

In the context of pandemic-related data analytics, data may, for example, be provided by various means, e.g., hand-filled forms, data provided through official institutions (hospitals, government agencies, etc.), data being extracted from automated processing, etc. Also, data originates typically from different sources, e.g., private persons, public services, not-for-profit organizations, etc. In some approaches, the data ingestion functional block is complemented by "data normalization functional blocks," which create data sets of a specific common format.

The functional blocks of ingestion and data normalization are typically fed by the "knowledge management functional block". A knowledge management entity is indeed fundamental in the context of Artificial Intelligence (AI). Typically, knowledge representation defines a set of formalisms that define information and knowledge in a computer-readable form. Formal and consensual knowledge representation also enables machine learning and reasoning. To extend the knowledge base of the system, an inference system is being applied.

In the context of pandemic-related applications, suitable hypotheses may relate to a refined classification of data sources to identify typical patterns characterizing the spreading of a disease, etc.

The "context-aware management functional block" enables the system to gather information about itself and its environment. Thus, finally, the system adapts its behavior according to changes in the context. Indeed, policy decisions are typically driven by the contextual history of a user (or data source), etc. In the context of a pandemic, information may be included related to a lockdown (confinement) of the population, movement restrictions, and other protective measures (e.g., wearing of masks), etc. Contextual data typically includes personal and group information of a user or data source (such as contact information, profile and preference information, etc.) etc.

In the context of public services, the sources of information may be typically classified in terms of (home) address information, the professional context of a data source, ownership of assets, etc.

The "cognition framework functional block" targets to imitate the function of the human brain to understand concepts; this is achieved using a set of specialized data structures and computational procedures operating similarly to the human brain. Cognition is used to process new data, apply inferences, and compare the results to available knowledge. Typically, a cognition entity comprises at least the following three functions:

i. Interfaces interacting with the environment providing data;
ii. Processing that can analyze and manipulate data, information, and knowledge;
iii. Storage to hold data, information, and knowledge.

In practice, cognitive systems are built based on one of the following two approaches: The symbolic approach (computationalism) and a connectionist approach, assuming that all cognitive processes are the same and derived from neural activity dynamics. Knowledge is obtained through the processing of training samples.

In the context of pandemic-related applications, a connectionist approach may be used to feed a model with existing data, indicating which samples imply specific patterns of spreading of a disease, persons not obeying medical requirements (e.g., confinement, wearing of masks, etc.), etc.

The "situational awareness functional block" enables the system to understand what has just happened, what is likely to happen, and how both may affect the objectives of the system. Typically, the process comprises five actions: gathering data (perception), understanding the significance of the data (through facts and inferences), determining what to do in response to the given event, making a decision, and executing those actions.

In the context of pandemic-related applications, situational awareness may relate to the current situation of a user (data source) and anticipated follow-up events, such as further spreading of a disease, contamination of new regions, etc.

The "model-driven engineering functional block" is responsible for enabling software development based on models instead of code. Related models are machine-readable. In our context, the use of such models fulfills three key purposes:

i. The model ensures that all different data models maintain a consistent definition and understanding of concepts;
ii. The model enables different policies at different levels of abstraction to communicate with each other using a common vocabulary and data dictionary;
iii. The model decouples the need for policy (defined by the situational awareness functional block) from the specification of policy from the implementation of policy.

In the context of the pandemic related application, the model-driven engineering approach is applying models suitable to characterize the spreading of a disease, behavioral information of a population, medical coverage and saturation level of the medical system, etc. and is thus able to accelerate the development of new functions and software applications.

Finally, the "policy management functional block" provides a set of uniform and intuitive mechanisms for consistent recommendations and commands meeting the following requirements:

i. To be able to transform data and information to a common format that facilitates generating outputs;
ii. To use a set of models including data types and structures for producing outputs;
iii. Data Denormalization may be achieved by a separate functional block.

In general, the policy is a natural way to express rules and restrictions on behavior and then automate the enforcement of those rules.

In the context of pandemic-related applications, a policy management framework will enable non-experts to apply a data evaluation system by providing evaluation requirements and objections through easy-to-read policies.

5 Managing Data to Solve the Modelization Problem

As described in the previous section, the Vaccine Allocation Algorithm (VAA) requires a large amount of data obtained from national authorities, local vaccination sites, and patients. The aggregation of this information in a single location causes many problems in terms of security, storage, data flows, etc. Thus, the authors propose introducing a Federated Data Management system that relies on Permissioned Distributed Ledger (PDL) for data storage (ETSI, 2021). This proposal is introduced her to secure and manage data in a distributed manner and adjust the vaccination problem to obtain a Distributed Vaccine Allocation Algorithm (DVAA) (Fig. 6).

a- **Description of the ETSI proposal (ETSI, 2021)**

b- **Introduction of Permissioned Distributed Ledger (PDL) in the Vaccine Allocation Model**

This section describes how to implement a PDL by introducing authorized nodes to ensure data circulation and enable to implementation of the VAA (Fig. 7). The architecture described in the following figure relies on explanations given in Table 1.

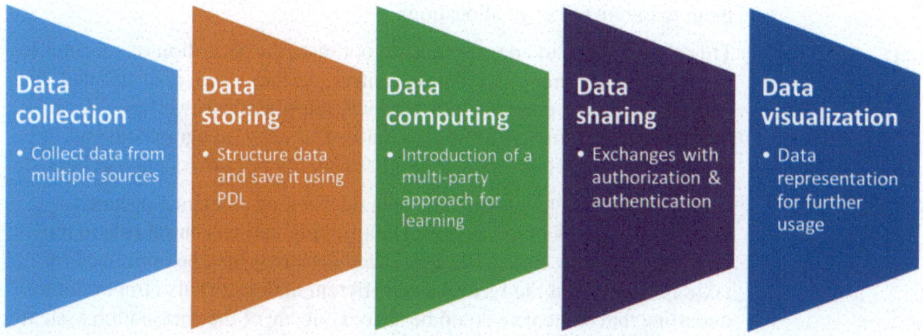

Fig. 6 Modules introduced to ensure Federated Data Management

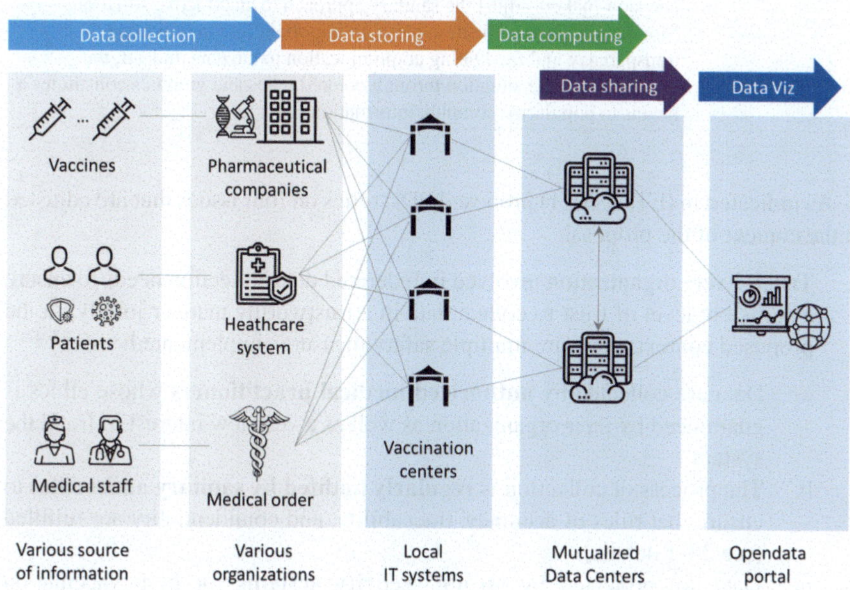

Fig. 7 Permissioned Distributed Ledger (PDL) in the Vaccine Allocation Model

Table 1 Correspondence between modules and the vaccination model

Module	Correspondence in the vaccination model
Data collection	This module aims to obtain data from every node in the system. This corresponds to data from patients (age, address, comorbidities, vaccination status, ...), from vaccines (pharmaceutical laboratory, available doses, stock identifier, ...), from vaccination centers (location, area, vaccination rooms, the flow of patients, ...) and medical practitioners (job specialty, availability, vaccination center assignment, ...)
Data storing	This module aims to ensure that data is saved in a structured manner so that data may be combined on-demand. The introduction of Permissioned Distributed Ledgers aims to ensure that data is secured by authorized nodes of the system. The interest of the distributed approach is to facilitate the exchange of data between nodes and to offer them the capability to combine them to obtain effective allocations
Data computing	This module aims to compute data to optimize the allocation of vaccines to people in the different vaccination centers. As the system is distributed, the algorithms should take into account local information as well as coordination parameters to fulfill a defined strategy (geographical and inter-generation orientations, as well as equality or fairness approaches)
Data sharing	This module aims to ensure that some decision parameters are shared between the actors of the system. Data sharing plays a central role in the optimization along time as decentralized decisions may be optimized by taking into account the feedback of different nodes. It is also important to underline that some data could be shared outside of the vaccination system to ensure a reasonable level of transparency to the citizens. Indeed, the limit is defined by protecting people's rights to privacy
Data visualization	This module aims to represent data through different software to help decision-makers adjust the sanitary approach to an effective situation. Moreover, data visualization plays an important role in providing transparency and facilitating communication to citizens. Indeed, the illustration of the situation through colored maps and graphics constitutes a means to popularize complex information

As indicated in (ETSI, 2021) proposed PDL relies on four issues that are adjusted in the context of the proposal:

1. **Trust**. Every organization involved in federated data collection needs to ensure a sufficient level of trust to collect data in a trustworthy manner jointly. In the proposed context, there are multiple safety measures implemented:
    a. Data are collected by **authorized medical practitioners** whose ethics is guaranteed by their organization as well as a very low interest to fraud the system
    b. The process of collection is **regularly audited by sanitary authorities** to ensure that rules of accuracy, traceability, and confidentiality are fulfilled (see 2 for details)
    c. Data are protected by **strong security systems** for both vaccination certificates and data exchanges (see 4 for details)

2. **Integrity**. Every node has to ensure that data may not be modified or corrupted. A combination of various methods may be employed to fulfill this objective.
   a. The usage of **distributed storage and local secret sharing** inspired from (Kim et al., 2019) may protect data from corruption while ensuring a limited impact on communication costs between nodes.
   b. The criticality of the healthcare system may imply relying on **two complementary storage systems**. Indeed, the classic IT central system could be supplemented by an "emergency PDL system" that communicates with each other. Although the combination of these two systems may appear heavy and costly, the government spending when a pandemic occurs afford to benefit from the combination of security and flexibility.
3. **Participation**. The participation of each node may rely on different approaches.
   a. The classic PDL approach consists in introducing an **incentive** that can rely on a **financial reward** on each data obtained after a vaccination occurs. The payment may be associated with the vaccination process or distinct depending on the process retained by the national healthcare institution. This payment relies on a fixed amount which ensures equality or on a degressive reward to support smaller or distant nodes (vaccination centers should cover the whole territory, including rural zones).
   b. As the pandemic is severe and harshly strikes the economy, governments may decide to set **mandatory and free participation** to the PDL. The PDL contributes to the vital interests of the nation. Furthermore, this approach reveals the easiest to implement as there is no monetization to define.
4. **Security**. Every node and interconnection of the system should ensure a high level of security as health data represent a special category in the GDPR..[2]
   a. Each stored data should be encrypted using a **Public Key Infrastructure** as proposed for the eHealth Network defined by the European Commission.[3] In France, the cyphering may be ensured using NIST P-521 as defined on page 26 of (ANTS, 2021) (Fig. 8).

6 Conclusion and Perspectives

In the present study, a novel approach is introduced, which combines Artificial Intelligence-based tools with classical, Lagrangian convex optimization techniques. Artificial Intelligence is used to extract input and boundary information from all available sources, then applied to the convex optimization step. As a result, an optimum vaccination strategy is derived serving well-defined objectives (minimizing

[2] GDPR: General Data Protection Regulation, https://eur-lex.europa.eu/eli/reg/2016/679/oj.

[3] Details available at *Detailed technical specifications for EU Digital COVID Certificates*, https://ec.europa.eu/health/ehealth/covid-19_en.

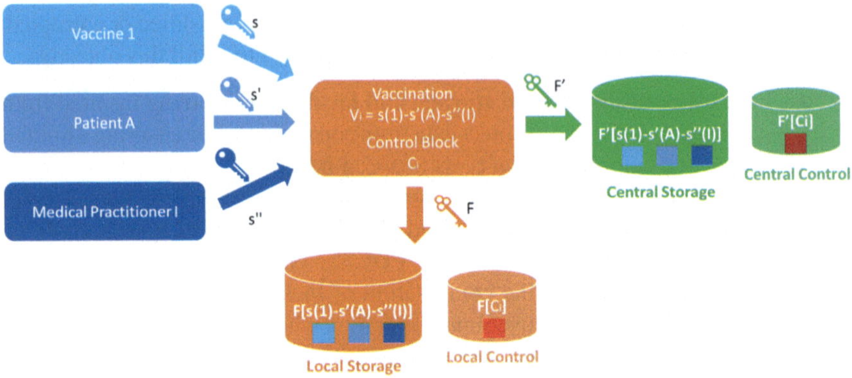

Fig. 8 Data security within the Permissioned Distributed Ledger

the number of deaths, achieving a tradeoff with economic objectives, etc.) and meeting boundary conditions, such as regional fairness, essential needs of citizens, etc. Finally, it is discussed how to ensure that reliable and trustworthy data is being used as a basis for the optimization process—including the possibility of validating the integrity of the data, verification of the source, etc. A solution is detailed, which relies on Permissioned Distributed Ledger technology, meeting the identified requirements. In summary, an end-to-end solution is proposed for a reliable derivation of an optimum vaccination strategy.

References

Abo, S., & Smith, S. R. (2020). Is a COVID-19 vaccine likely to make things worse? *Vaccines., 8*(4), 761. https://doi.org/10.3390/vaccines8040761.

Alankar, B., Kaur, H., Ahsaan, S., Sharma, G., & Chang, V. (2021). Towards reviewing an immediate impact of COVID-19 on the integrative world economy: An evolving perspective. *Journal of Global Information Management (JGIM), 30*(4), 1–19.

Antal, C., Cioara, T., Antal, M., & Anghel, I. (2021). Blockchain platform For COVID-19 vaccine supply management. *IEEE Open Journal of the Computer Society, 2*(01), 164–178. https://doi.org/10.1109/OJCS.2021.3067450.

Arani, S., Nawab, M., Rahman, M., & Zaman, M. (2020). A Blockchain-based approach to prevent hidden contagion of COVID-19. *Compiler, 9*(2), 71–84. https://doi.org/10.28989/compiler.v9i2.787.

Boyd, S., & Vandenberghe, L. (2004). *Convex optimization*. Cambridge University Press.

Bubar, K. M., Reinholt, K., Kissler, S. M., Lipsitch, M., Cobey, S., Grad, Y. H., Larremore, D. B. (2021, February). Model-informed COVID-19 vaccine prioritization strategies by age and serostatus. *Science, 26;371*(6532), 916–921. https://doi.org/10.1126/science.abe6959; Epub 2021 Jan 21. PMID: 33479118; PMCID: PMC7963218.

Buhat, C. A. H., Lutero, D. S. M., Olave, Y. H. et al. (2021). Using constrained optimization for the allocation of COVID-19 vaccines in the philippines. *Applied Health Economics and Health Policy, 19*, 699–708. https://doi.org/10.1007/s40258-021-00667-z.

Carniel, A., Leme, G., Bezerra, J., & Hirata, C. (2021). A Blockchain approach to support vaccination process in a country. In *Proceedings of the 23rd International Conference on Enterprise Information Systems* (Vol. 1, pp. 343–350): ICEIS, ISBN 978-989-758-509-8. https://doi.org/10.5220/0010520003430350.

Dhiman, G., Chang, V., Kant Singh, K., & Shankar, A. (2021). Adopt: Automatic deep learning and optimization-based approach for detection of novel coronavirus covid-19 disease using x-ray images. *Journal of Biomolecular Structure and Dynamics* 1–13.

Estiri, H., Strasser, Z.H., Klann, J.G. et al. (2021). Predicting COVID-19 mortality with electronic medical records. *NPJ Digital Medicine* **4**, 15. https://doi.org/10.1038/s41746-021-00383-x.

ETSI (Sept 2019). Group Specification (GS) ENI 005, V1.1.1. http://www.etsi.org.

ETSI GR PDL009 (2021, September 03). Permissioned distributed ledger for federated data management. https://www.etsi.org/deliver/etsi_gr/PDL/001_099/009/01.01.01_60/gr_PDL009v010101p.pdf.

Gaie, C., & Mueck, M. (2022). An artificial intelligence framework to ensure a tradeoff between sanitary and economic perspectives during the COVID-19 pandemic, deep learning for medical applications with unique data.

Gaie, C. (2010). Allocation de resources à prise de décisions distribuées dans des réseaux mobiles hétérogènes. Ph.D. dissertation, Facult´edessci-ences d'Orsay, Universit´e Paris-Sud 11, April, 2010.

Gaie, C., Assaad, M., Muck, M., & Duhamel, P. (2008). Distributed discrete resource optimization in heterogeneous networks. In *IEEE Workshop on Signal Processing Advances for Wireless Communications (SPAWC 2008)*. https://doi.org/10.1109/SPAWC.2008.4641670.

Gupta, M., Kumar, V., Yadav, V., Singh, R. K., & Sadim, M. (2021). Proposed framework for dealing COVID-19 pandemic using blockchain technology. *Journal of Scientific and Industrial Research, NISCAIR-CSIR, India, 80*(3), 270–275. ISSN 0975-1084.

Gong, J., Ou, J., Qiu, X., Jie, Y., Chen, Y., Yuan, L., Cao, J., Tan, M., Xu, W., Zheng, F. Shi, Y., & Hu, B. (2020, Augusr). A tool for early prediction of severe coronavirus disease 2019 (COVID-19): A multicenter study using the risk nomogram in Wuhan and Guangdong, China, Clinical infectious diseases. **71**(15.1), 833–840. https://doi.org/10.1093/cid/ciaa443.

Han, S., Cai, J., Yang, J., et al. (2021). Time-varying optimization of COVID-19 vaccine prioritization in the context of limited vaccination capacity. *Nature Communications, 12*, 4673. https://doi.org/10.1038/s41467-021-24872-5.

Jabarulla, M. Y., & Lee, H.-N. (2021). A blockchain and artificial intelligence-based, patient-centric healthcare system for combating the COVID-19 pandemic: opportunities and applications. *Healthcare, 9*(8), 1019. https://doi.org/10.3390/healthcare9081019.

Kaur, H., Ahsaan, S. U., Alankar, B., & Chang, V. (2021). A Proposed sentiment analysis deep learning algorithm for analyzing COVID-19 tweets. *Information Systems Frontiers, 23*, 1417–1429.

Kim, J. H., Marks, F., & Clemens, J. D. (2021). Looking beyond COVID-19 vaccine phase 3 trials. *Nature Medicine, 27*, 205–211. https://doi.org/10.1038/s41591-021-01230-y.

Kim, Y., Raman, R. K., Kim, Y., Varshney, L. R., & Shanbhag, N. R. (2019, February). Efficient local secret sharing for distributed blockchain systems. *IEEE Communications Letters, 23*(2), 282–285. https://doi.org/10.1109/LCOMM.2018.2886016.

Leng, A., Maitland, E., Wang, S., Nicholas, S., Liu, R., & Wang, J. (2020). Individual preferences for COVID-19 vaccination in China. *Vaccine.* 2021 **8, 39**(2), 247–254. https://doi.org/10.1016/j.vaccine.2020.12.009; Epub 2020 Dec 5. PMID: 33328140; PMCID: PMC7719001.

Matrajt, L., Eaton, J., Leung, T., Dimitrov, D., Schiffer, J. T., Swan, D. A., & Janes, H. (2021). Optimizing vaccine allocation for COVID-19 vaccines shows the potential role of single-dose vaccination. *Nature Communications, 12*, 1. https://doi.org/10.1038/s41467-021-23761-1.

Matrajt, L., Eaton, J., Leung, T., & Brown, E. R. (2020). Vaccine optimization for COVID-19, who to vaccinate first? medRxiv: the preprint server for health sciences. https://doi.org/10.1101/2020.08.14.20175257.

Mcdonnell, A., Exan, R. V., Chalkidou, K., Lloyd, S., Subramanian, L., Rosenberg, J., Weintraub, R., Porta, A., Li, J., Reader, D., Maiza, E., Scannell, J., Thomas, V., & Yadav, P. (2020). COVID-19 vaccine predictions: using mathematical modelling and expert opinions to estimate timelines and probabilities of success of COVID-19 vaccines.

Rauchs, M., Glidden, A., Gordon, B., Pieters, G. C., Recanatini, M., Rostand, F., Vagneur, K., & Zhang, B. Z. (2018, August 13). Distributed ledger technology systems: A conceptual framework. https://doi.org/10.2139/ssrn.3230013.

Suhasini, M., & Singh, D. (2021) Blockchain based framework for secure data management in healthcare information systems. *Annals of the Romanian Society for Cell Biology, 25*(6), 16933–16946. Retrieved from https://www.annalsofrscb.ro/index.php/journal/article/view/8985.

Schwarzinger, M., Watson, V., Arwidson, P., Alla, F., & Luchini, S. (2021). COVID-19 vaccine hesitancy in a representative working-age population in France: A survey experiment based on vaccine characteristics. *Lancet Public Health, Elsevier, 6*(4), e210–e221. https://doi.org/10.1016/s2468-2667(21)00012-8.

Singh, A., Jindal, V., Sandhu, R., & Chang, V. (2021). A scalable framework for smart COVID surveillance in the workplace using Deep Neural Networks and cloud computing. *Expert Systems* e12704.

Spécifications techniques des Codes à Barres 2D-Doc. (2021, April 30). Agence nationale des titres sécurisés https://ants.gouv.fr/content/download/516/5665/version/12/file/Specifications-techniques-des-codes-barres_2D-Doc_v3.1.3-02082021.pdf.

Tai, Y., Gao, B., Li, Q., Yu, Z., Zhu, C., & Chang, V. (2021). Trustworthy and intelligent covid-19 diagnostic iomt through xr and deep learning-based clinic data access. *IEEE Internet of Things Journal, 8*(21), 15965–15976.

Tang, B., Zhang, X., Li, Q. et al. (2021). The minimal COVID-19 vaccination coverage and efficacy to compensate for potential increase of transmission contacts, and increased transmission probability of the emerging strains, 14 January 2021, PREPRINT (Version 1) available at Research Square. https://doi.org/10.21203/rs.3.rs-140717/v1.

Correction to: Uncertainty Propagation and Salient Features Maps in Deep Learning Architectures for Supporting Covid-19 Diagnosis

Iñaki Rodríguez-García, Tomás Sánchez-Pastor, Julia Vázquez-Escobar, Juan Luis Gómez-González, and Miguel Cárdenas-Montes

Correction to:
Chapter "Uncertainty Propagation and Salient Features Maps in Deep Learning Architectures for Supporting Covid-19 Diagnosis" in: V. Chang et al. (eds.), *Artificial Intelligence and Machine Learning Methods in COVID-19 and Related Health Diseases*, Studies in Computational Intelligence 1023,
https://doi.org/10.1007/978-3-031-04597-4_1

In the original version of the chapter, the following belated corrections have been incorporated:

The author name "Tomás Sáchez-Pastor" has been changed to "Tomás Sánchez-Pastor" in the Frontmatter and in Chapter "Uncertainty Propagation and Salient Features Maps in Deep Learning Architectures for Supporting Covid-19 Diagnosis".

The chapter and the book have been updated with the change.

The updated version of this chapter can be found at
https://doi.org/10.1007/978-3-031-04597-4_1

V. Chang et al. (eds.), *Artificial Intelligence and Machine Learning Methods in COVID-19 and Related Health Diseases*, Studies in Computational Intelligence 1023,
https://doi.org/10.1007/978-3-031-04597-4_12

The manufacturer's authorised representative in the EU is Springer Nature Customer Service Centre GmbH, Europaplatz 3, 69115 Heidelberg, Germany. If you have any concerns regarding our products, please contact ProductSafety@springernature.com

Printed and bound by CPI Group (UK) Ltd, Croydon, CR0 4YY
22/09/2026
02221055-0007